Why do animals play? Play has been described in animals as diverse as reptiles, birds and mammals, so what benefits does it provide and how did it evolve? Careful, quantitative studies of social, locomotor and object play behavior are now beginning to answer these questions and to shed light on many aspects of both animal and human behavior.

This unique interdisciplinary volume brings together the major findings about play in a wide range of species including humans. Topics include the evolutionary history of play, play structure, function and development, and sex and individual differences. *Animal Play* is destined to become the benchmark volume in this subject for many years to come, and will provide a source of inspiration and understanding for students and researchers in behavioral biology, neurobiology, psychology, anthropology and behavioral medicine.

Animal Play:
Evolutionary, Comparative, and Ecological Perspectives

Animal Play: Evolutionary, Comparative, and Ecological Perspectives

Edited by

MARC BEKOFF
and
JOHN A. BYERS

PUBLISHED BY THE PRESS SYNDICATE OF THE UNIVERSITY OF CAMBRIDGE
The Pitt Building, Trumpington Street, Cambridge CB2 1RP, United Kingdom

CAMBRIDGE UNIVERSITY PRESS
The Edinburgh Building, Cambridge CB2 2RU, United Kingdom
40 West 20th Street, New York, NY 10011-4211, USA
10 Stamford Road, Oakleigh, Melbourne 3166, Australia

First published 1998

Typeset in 10 on 13pt Times [KW]

A catalogue record for this book is available from the British Library

Library of Congress Cataloguing in Publication data
Animal play: evolutionary, comparative, and ecological prespectives /
edited by Marc Bekoff and John A. Byers.
p. cm.
ISBN 0 521 58383 7 (hardbound) 0 521 58656 9 (pbk)
1. Play behavior in animals. I. Bekoff, Marc. II. Byers, John A., (John Alexander), 1948– .
QL763.5.A54 1998
591.56′3–dc21 97-22056 CIP

ISBN 0 521 58383 7 hardback
ISBN 0 521 58656 9 paperback

Transferred to digital printing 2004

Marc Bekoff dedicates this book to his parents, who always encouraged and continue to encourage play in their children

John A. Byers dedicates this book to his children, Anna and John, who have graciously reminded him of those parts of human play that he had forgotten

Contents

Contributors

Allen, Colin,
Department of Philosophy, Texas A&M University, College Station, Texas 77843 USA <colin-allen@tamu.edu>

Bekoff, Marc,
Department of Enviromental, Population and Organismic Biology, University of Colorado, Boulder, Colorado 80309-0334 USA <marc.bekoff@colorado.edu>

Biben, Maxeen,
2014 Martinsburg Pike, Winchester, Virginia 22603 USA

Brown, Stuart,
225 Crossroads Blvd. Box 341, Carmel, California 93923 USA <Stuplay@authenticplay.com>

Burghardt, Gordon M.,
Department of Psychology, University of Tennessee, Knoxville, Tennessee 37996 USA <gburghar@utk.edu>

Byers, John A.,
Department of Biological Sciences, University of Idaho, Moscow, Idaho 83844-3051 USA <jbyers@uidaho.edu>

Hall, Sarah L.,
Anthrozoology
Institute, University of Southampton, Basset Crecent East, Southampton SO16 7PX, UK

Heinrich, Bernd,
University of Vermont, 111 Marsh Life Science Building, Burlington, Vermont 05405 USA

Miller, Michelle N.,
Biology Department, Blue Mountain Community College, Pendleton, Oregon 97801 USA <mmoodie@bmcc.cc.or.us>

Pellis, Sergio M.,
Department of Psychology and Neuroscience, University of Lethbridge, Lethbridge, Alberta T1K 3M4 Canada <*pellis@hg.uleth.ca*>

Pellis, Vivien C.,
Department of Psychology and Neuroscience, University of Lethbridge, Lethbridge, Alberta T1K 3M4 Canada

Siviy, Stephen M.,
Department of Psychology, Box 407, Gettysburg College, Gettysburg, Pennsylvania 17325-1486 USA <*ssiviy@gettysburg.edu*>

Smolker, Rachel,
University of Vermont, 111 Marsh Life Science Building, Burlington, Vermont 05405 USA <*rsmolker@zoo.uvm.edu*>

Thompson, Katerina,
Biological Sciences Program, College of Life Sciences, University of maryland, College Park, maryland 20742 USA <*KACI@life.umd.edu*>

Watson, Duncan M.,
School of Horticulture, University of Western Sydney, Hawkesbury, Richmond, New South Wales, Australia, 2753 <*D.Watson@uws.edu.au*>

Introduction

Like many other beautiful things, play is ephemeral, and often difficult to study. Play typically occupies a small part of an animal's day, and in most species, occurs only during a circumscribed part of juvenile life. Other types of behavior are more approachable and more forgiving. To study foraging in pronghorn, one can show up at just about any time of day. Missing a week or so of data is not a problem: one can always get more samples, because the animals are always foraging. In contrast, to study play in this (and other) species, one must be prepared to start observations at 5:00 a.m. each day. A missing week means that about 15% of the behavior is missed. Behavioral development is always more difficult to study than adult behavior, and play is more difficult to study than most other aspects of behavioral development. Nevertheless, play is receiving increased attention from many students of behavior studying an increasing number of different taxa, and interdisciplinary efforts are beginning to pay off. It also is becoming clear that a detailed understanding of play allows for strong insights into other aspects of behavior and social ecology. This book, which grew out of a symposium organized by Marc Bekoff and Robert Fagen for the 1996 Animal Behavior Society meetings in Flagstaff, Arizona, presents the thoughts and recent empirical results of many of the leaders in play research. We have not broken down the volume into distinct sections because of extensive overlap among many of the essays, overlap that compels researchers to continue broadly comparative and evolutionary research in this exciting area.

The category of behavior called play has challenged students of behaviour for a long time. For some, play was written off as behavior not worth studying. For others, the difficulty of studying play rigorously prompted the claim that a sufficiently 'scientific' study of play was impossible. Others, recognizing that play was not only interesting and beautiful but also essential to understand, rose to the challenge and began detailed observational and experimental work. An early symposium organized by Marc Bekoff and published in the *American Zoologist* in 1974 contains several now-classic papers, including one in which Susan Wilson and Devra Kleiman introduced the term 'locomotor-rotational play'. The field entered its modern phase in 1981 with the publication of Robert Fagen's now classic *Animal Play Behaviour*. Fagen provided a comprehensive review of the phylogenetic distribution of play, and outlined the major avenues of research that are under investigation today. This book is centred on five themes, four of which were discussed by Fagen.

Many of the unanswered questions about play that Fagen posed are still unanswered. The most basic of these is, 'which animals play?' Incredibly, we still do not know if most mammals do or do not play, and our knowledge of the distribution of play in other vertebrate classes is weaker still. This is an unfortunate state of affairs, because the phylogenetic distribution of a trait provides fundamental evidence about its origin, evolutionary history, and possible functions. Three chapters in the book offer new data and challenges in this area. Burghardt (Chapter 1) shows that play may occur in some reptiles, and raises the vital issue of how one might recognize play in its rudimentary, or ancestral stage. Heinrich & Smolker (Chapter 2) show that elaborate social and object play occur in ravens. Watson (Chapter 4) provides a comprehensive review of play in the Macropodoidea, the marsupial ungulate equivalents of Australia. These essays show that researchers must go beyond primates and other well-known mammals in their comparative studies.

A second basic but unanswered question is 'what are the typical age-specific rates of play?' Byers (Chapter 10) points out that good descriptive information on this question is surprisingly scarce. He also argues that the age distribution of play is an essential clue about its true biological effects. We will be in a much better position to ask the right questions about play when we have a clear picture of many species-typical age-specific play rates. In addition to the typical juvenile expression of play in many mammals, some forms of play persist beyond sexual maturity. Hall (Chapter 3), Bekoff & Allen (Chapter 5), Miller & Byers (Chapter 7), and Biben (Chapter 8) provide examples of play and play-like behaviour in adults.

A third question about play, unanswered despite Fagen's (1981) emphasis and much previous and subsequent speculation, concerns biological effects or function. What exactly does play do for a young animal? What benefits outweigh the very real energy and risk costs of play?

Although many agree that play seems to have something to do with motor development, or development of cognitive skills that support motor performance (Biben, Chapter 8; Siviy, Chapter 11), hard evidence is scant and opinions are divided. Pellis & Pellis (Chapter 6) show that several details of rat play fighting are not consistent with motor training for fighting skills, but Heinrich & Smolker (Chapter 2), Miller & Byers (Chapter 7), Biben (Chapter 8), and Thompson (Chapter 9) show in various ways that motor training or cognitive/motor training are likely in ravens, pronghorn, squirrel monkeys, and humans. Watson (Chapter 4) shows that most, but not all, aspects of kangaroo play fighting match the predictions of the motor training hypothesis. Byers (Chapter 10) arguing that the motor training hypothesis is far too broad, suggests a method to refine it, and presents evidence against the popular notion that play is a way of getting into shape. A theme that emerges from all chapters that discuss function is the need to refine hypotheses. There also is a need for more detailed descriptions of the age distribution of play. More detailed descriptions of the motor acts used in play will also be useful. Pellis & Pellis (Chapter 6) show that a closer look at rat play fighting, documenting movement by movement orientations and targets of attack reveals that play fighting resembles adult fighting in only some ways, and resembles adult sexual behavior in others. This analysis and others, such as Biben's (Chapter 8) and Thompson's (Chapter 9) raise the following question: just how closely must motor act 'A' resemble motor act 'B' before we conclude that 'A' can serve as practice, or motor training for 'B'? This is an unanswered question over which opinion is currently divided.

A fourth theme of this book concerns the mechanisms that produce and maintain play. As Hall (Chapter 3) correctly notes, 'Study of play behaviour suffers from a neglect of non-evolutionary causes. Very little emphasis is placed on proximate causes and mechanisms'. As is so for the study of many aspects of behavior, study of the proximate mechanisms that control play has lagged far behind the study of structure and function. Hall describes experiments designed to test the hypothesis that play and predatory aggression share a common motivation. Siviy (Chapter 11) demonstrates how pharmacological techniques may be used to study the neurotransmitter systems involved in play, and how recently developed neural imaging techniques may be used to identify the areas of the brain that are most active when animals play.

The fifth theme of this book concerns the communicative and cognitive aspects of play. Biben (Chapter 8) stresses the importance of developmental flexibility in the social behavior of rhesus monkeys, and suggests that animals use play as a way to learn strategies of social interaction. Bekoff & Allen (Chapter 5) point out that the strongly cooperative nature of much social play makes it a useful model for study of the communication of intention. For example, when two animals engage in play fighting, there is a real risk of injury to each, especially if one switches from play fighting to real fighting (Miller & Byers, Chapter 7; Biben, Chapter 8). Somehow, the playful and cooperative nature of the interaction is maintained, even though the motor acts themselves may closely resemble aggression. How do animals 'read' play intention in a conspecific? Cooperative social play may involve rapid exchange of information on intentions, desires, and beliefs. Discussions of self-handicapping and role-reversals in play by individuals of many taxa can also inform cognitive inquiries. Finally, for individuals of species in which cooperative social play is a form of social cognitive training, what are the consequences of failure to play? Brown (Chapter 12) suggests, with a fascinating set of correlations, that play in childhood is required for complete social integration of humans.

The five themes of this book come and go throughout each chapter, flickering across treatment of this elusive and beautiful behavior much in the same way that play itself flickers across the development of an individual. We hope to inspire research in each of the five themes; there is still a lot of work to be done. Foremost, we hope that you will enjoy this book and will come to share our own long-time fascination with play.

Marc Bekoff thanks the Animal Behavior Society for providing partial support for the above-mentioned symposium, and also Robert Fagen for his help in organizing the gathering. We both thank Tracey Sanderson, Cambridge University Press, for her support for this project.

Boulder, Colorado *Marc Bekoff*
Moscow, Idaho *John A. Byers*

1

The evolutionary origins of play revisited: lessons from turtles

GORDON M. BURGHARDT
Departments of Psychology and Ecology & Evolutionary Biology, University of Tennessee, Knoxville, TN 37996-0900, USA.

The current status of identifying nonavian reptile play

The origins of vertebrate play are obscure, but more understanding of these origins would aid greatly in clarifying and evaluating hypotheses about play (Bekoff & Byers 1981; Burghardt 1984, 1998). It has been clearly shown that some birds and most if not all groups of mammals show behavior currently classified as play (Ficken 1977; Fagen 1981; Ortega & Bekoff 1987). Is play behavior restricted to endothermic vertebrates with extensive parental care? Since the nineteenth century, during which claims for play were made for many invertebrates and vertebrates (Fagen 1981; Burghardt in press; Bekoff & Allen, Chapter 5), the generally accepted phyletic scope of play has become narrowed to the extent that it is generally limited to mammals and birds. If credible evidence for play outside of the mammalian and avian radiations is to be sought, key groups are found in the nonavian reptiles. Although I will use the term reptiles from here on, many authorities hold that reptiles are not a monophyletic group, that share a common ancestor. However, even if reptiles are monophyletic, birds would be part of that group, related most closely to crocodilians. Crocodilians are in many physiological, paleontological, and life history characteristics more similar to birds than other traditional reptile groups such as turtles. For example, crocodilians have a four-chambered heart and all show postnatal parental care, complete with a complex vocal communication system that includes 'contact' and 'distress' calls (Herzog & Burghardt 1977). As archosaurs, they share a more recent common ancestor with birds than with squamate reptiles or turtles.

Until recently, the lack of unequivocal documented evidence of replicable play like behavior in nonavian reptiles supported the view that play was not found at this grade of organization. In an exhaustive survey of the literature on animal play, Fagen (1981) could locate only two purported examples of reptile play: anecdotal reports of object play in a captive Komodo monitor lizard (*Varanus komodoensis*, Hill 1946) and a field observation of an American alligator (*Alligator mississippiensis*) repeatedly attracted to, and snapping at, water dripping from a spout (Lazell & Spitzer 1977). Unfortunately, limited details and no filmed records are available for these observations and confirming examples have not been published. Anecdotal claims of play in reptiles are sometimes found in the pet literature (e.g., green iguanas, *Iguana iguana*, Hatfield 1996). Play has been invoked as a possible explanation for enigmatic neonatal behavior such as nonsocial head bob displays in fence lizards (*Sceloropus undulatus*, Roggenbuck & Jenssen 1986) and 'wrestling' in African chameleons (*Chamealo bitaenlatus*, Burghardt 1982). Since systematic description and experimental analysis was not available for these examples, the existence of play in nonavian reptiles remained in doubt. As reviewed below, several life history, ontogenetic, physiological, and psychological factors mitigate against, but do not preclude, the occurrence of 'typical' mammalian playfulness in nonavian, ectothermic reptiles (Burghardt 1984, 1988).

In this chapter I review recent data suggesting that object, locomotory, and social play do occur in some turtles, the oldest group of extant reptiles and closest to the basal therapsid stock. This is accomplished first by describing phenomena that, if seen in a mammal or bird, would be readily labeled playful by most observers. Later I compare in more detail these examples with frequently noted criteria for play. These examples may aid in identifying the *primary* processes by which 'play' originated and evolved in ancient vertebrates and their modern descendants. Understanding the shadowy origins of play may provide the essential scaffolding from which the highly diverse and complex structures of mammalian and avian play could evolve through a series of *secondary* or derived processes. Indeed, playfulness, largely ignored in evolutionary approaches to animal learning and cognition, social organization, foraging ecology, niche expansion, and behavioral ontogeny, may be an important aspect of behavioral innovation in the evolution of vertebrates. For example, a recent interdisciplinary volume (Belew & Mitchell 1996) promoting the Baldwin Effect as a prime factor in evolution made no mention of play. The Baldwin Effect posits a process by which novel variations can become genetically fixed. Play could be a prime source for the novel behavior needed for the principle to work. Turtles may also be a key group to study for reasons other than mere comparative stamp collecting. Before presenting recent evidence on turtle play I will outline some of the lessons turtles might teach us.

Function and evolution are two distinct issues

According to Fagen (1981) early ethological research on animal play was often devoted to a careful description of the behavioral topography and contexts of play: the structural approach. In recent years, stimulated by the development of sociobiology (Fagen 1981), much research has been directed at evaluating the costs and benefits of playful behavior: the functional approach. Unfortunately, the focus on, and debates concerning, the structure–function dichotomy in studying play has not contributed substantially to our knowledge of the origins of play, either phylogenetically or ontogenetically. This result may be related to isolating studies of structure *or* function from each other at the expense of attention to the 'structure–function interface' (Pellis & Pellis, Chapter 6). The preoccupation with the structure–function dichotomy may also have led to the lack of clear conceptual distinctions between issues of origin and phylogeny from issues of function or current adaptiveness.

The functional approach of today is marked by a concern with the costs and benefits of play. In addition to the widespread popularity of economic models in behavioral ecology generally, there are other persuasive reasons for this emphasis on cost-benefit analysis in the study of play. One is the often noted observation that the lack of serious scientific attention to play is due to the fact that play is, well, not really a 'serious' (i.e., important) factor in behavior or ontogeny. This ties into a second reason, which is the difficulty researchers have had in demonstrating clear functions of play in animals (Martin & Caro 1985) and even humans (Smith 1996). In most areas of behavior, the functional approach has yielded great rewards rather quickly once adaptive explanations have been carefully stated and explored. Unfortunately this has not been the case with play, although this book presents some promising new leads. Perhaps for these reasons researchers have acted as if basic issues of origin, phylogeny, and even structure should take a back seat until play can be shown to be clearly beneficial in later life to be of practical importance in understanding major categories of behavior (e.g., foraging, fighting, courting), or even to be an important player in solving problems in child rearing, education, or crime (see Brown, Chapter 12). Perhaps, however, in play more than other areas, progress necessitates the integration of all four of the classical ethological aims: causation, function, ontogeny, and evolution (Tinbergen 1963).

Although the evolution of traits is based on their adaptedness and fitness consequences, demonstrating function, *per se*, tells us little about evolution or even the origin of the traits themselves. The current utility of playfulness in the lives of animals may actually mislead us about the contexts facilitating the evolution of play. It is these contexts that may be critical for a comprehensive understanding of the proximate causes and functions of play. Studies on the utility of play can help us develop hypotheses about the origins and phylogeny of playfulness but such studies can not substitute for a direct assault on the origins and phylogeny issue. In the language of modern comparative biology, the problem is that it is impossible to use only derived (apomorphic) characters, and the most spectacular ones at that, in a limited set of species to reconstruct the phylogeny and course of evolution of a trait. How do behavioral and other characters associated with play map onto a phylogeny? What traits are correlated and what is the nature of the correlation (Pagel 1994)? Eventually play research must deal with comparative issues (Martins 1996) much more rigorously than has been the case hitherto. Few studies on the evolution of play actually deal with the origins of playfulness.

Top down vs bottom up

Most scientists try to elucidate the meaning and function of animal play by studying the most clear, complex, and indisputable instances of play. This is understandable. Play is often a delightful and complex behavior that attracts human interest. Many companion animals serve largely as play partners for the child (or parent!) in us. Play, as Fagen has pointed out (e.g., Fagen 1981, 1996), provides an aesthetic experience for the human observer. That play may be a source of creative and artistic endeavors was suggested by Spencer & Schiller in the nineteenth century. Watching animals play is just plain fun. Who has not been enraptured by gibbons swinging from ropes in a large zoo enclosure, a kitten batting and pouncing on a soft ball, juvenile baboons playing chase games, or a dog and owner playing retrieve games in the park (Mitchell & Thompson 1991)?

Another reason for the focus on the dramatic and compelling in play echoes a point mentioned in the previous section. Since play is not yet in the mainstream of behavioral study, it might be most useful to concentrate on examples of play that appear to have the best promise of convincing skeptics that playfulness is deserving of serious attention.

Surely the most dramatic, complex, and time consuming kinds of play consume the most energy and must be the most adaptive. However, the history of science is littered with examples where the most obvious candidates did not prove to be the best choice for study. Who would have predicted twenty years ago that major breakthroughs in genetics, development, neural networks, and behavior would come from the study of a tiny nematode worm that merited virtually no mention in general biology courses a decade ago (Riddle et al. 1996)?

Thus, as valuable and satisfying as studying elaborate and compelling behavior may be, the overall value of such studies may be less than apparent until we understand much more about the biology and evolutionary origins of playfulness, the *primary* processes underlying play behavior. In 1995, when I attended the World Congress on play in Austria, I saw, at Vienna's Kunstmuseum, Breugel's famous sixteenth century painting of children playing many games. Although studying the many dozens of games and toys depicted in this painting may help us understand what sixteenth century Belgian children did, the profusion of play activities depicted would overwhelm any serious analysis of why children played and how those numerous games came to be in their repertoire. Many *secondary* processes have added so many layers of diverse complexity that the more basic underlying processes are obscured.

To more effectively understand the contexts facilitating playful behavior and its subsequent evolution as prominent, even critical, traits in animal life, a phyletic bottom up, rather than top down, approach is needed. We need to study 'play' in animals where it seems questionable, infrequent, subtle, and context dependent (Burghardt 1984), especially in outgroups and sister groups at all taxonomic levels. Furthermore, it is likely that more parsimonious explanations may be identified and evaluated in such cases than may seem called for in explaining elaborate and derived play. By answering questions such as what may be eliciting, controlling, precluding or limiting the occurrence, diversity, and richness of play in such animals, we may gain insights into why it is so prominent, complex, and even necessary in others. Schusterman & Gisiner (1997) have advocated a bottom up approach to animal cognition that, although different in emphasis, derives from a similar epistemological position.

Two perennial views on why play exists

Virtually all the explanations for the existence of play can be parceled into two basic views. These influence current approaches to play and the implicit reasons for its evolutionary pattern (Burghardt in press). The first, and by far most common, is that play is a preparation for the future. Certainly Groos (1898), with his practice theory, was a proponent of this view. Almost all functions of play are predicated on some benefit occurring later in life. Indeed, this delayed benefit view is so basic to virtually every definition of play that demonstrating a current utility to behavior labeled play is tantamount to proving that it is *not* play.

In contrast to the view that play is preparation for the future is the view that play is a legacy from the past. The recapitulation theory of G. S. Hall (1904) falls into this category. Hall saw the kicking, throwing, and hitting games of boys (in particular) as a vestige of the past in which these activities were important in hunting and warfare. Similarly, doll-playing in young girls could be interpreted as a vestige from a time when young girls were essential in helping to rear infant sisters. The surplus resource theory of Spencer (1872) also used the legacy of the past. Play was to be found in healthy well provisioned young of 'higher animals' where it was a blowing off of energetic steam in which the instincts already present were triggered. In this case any functions of play are derived. Whatever the problems of the legacy theorists, and they are great, they were on to a truth that is often pushed aside by the preparation theorists.

All play is not equal

One of the problems in defining, studying, and interpreting play is that it is most certainly a heterogeneous category. Although the traditional categories of locomotor, object, and social play are useful and are followed below, it is not clear whether all three kinds of play share more than superficial similarities. The fact that some animals, such as the black bears I once raised, may shift among all three play types in rapid succession or even perform them simultaneously does not prove that they have the same origins. Social play can, moreover, involve locomotor elements seen in solitary locomotor play and objects used in object play. Object and social play peak at different times in the behavior of young primates and carnivores (Fagen 1981) suggesting that they may have different causal bases, functions, phylogenies, and ontogenies that may secondarily coalesce. Animals that show only one type of play may allow a clearer analysis of the relevant primary processes.

The issue of reptile play

The presence of play in the reptile grade of organization is of critical interest from all the perspectives contrasted above. The controversy about the existence of play in reptiles raises issues of identification and documentation.

Many of the putative factors that may explain why play is rare or absent in reptiles were synthesized into the Surplus Resource Theory or SRT (Burghardt 1984). In brief, the theory has three main premises. First, except for crocodilians, most neonatal reptiles are not cared for by their parents and consequently must devote their activities toward surviving, avoiding predation, and growing rapidly on their own. As a result there is limited 'safe' time or opportunity for practicing or perfecting behaviors to be used in an uncertain future. Selection will have shaped abilities that enhanced juvenile survival, such as remaining quiet and inconspicuous. Nonavian reptiles are highly precocial and *superficially* resemble, and behave like, miniature adults. Second, reptiles typically are metabolically constrained from performing vigorous, energetically expensive behaviors not immediately beneficial (Bennett 1982). This is due to their low resting and maximal metabolic rate, limited aerobic capacity, and long recovery times after anaerobic expenditures. Taken together, these factors constrain the performance of and selection for, vigorous play like behavior with no current function. A third factor in the origins of play is psychological: animals in boring, unstimulating environments would be most likely to engage in behavior to relieve sensory and response deprivation and to increase arousal. Such boredom might be expected in the well provisioned and protected environments provided by endotherm parents.

As a guide to identifying behavior in nonavian reptiles that could be candidates for traditional appearing play, I subsequently (Burghardt 1988) predicted, using physiological, psychological, and life history processes, that mammalian or avian-like play in reptiles, whether locomotor, object, or social, should be rare and occur only in specific contexts in which those factors facilitating play in mammals and birds are also present.

Object play

Recently, we reported on detailed video-taped observations of object play in a Nile soft-shelled turtle, *Trionyx triunguis* (Burghardt et al. 1996).

This was a long-term (> 50 years) captive male, named Pigface, at the National Zoo in Washington, DC. This is a large species with a carapace length up to 95 cm and a mass of 90 kg (Ditmars 1933; Ernst & Barbour 1989). Recent phylogenetic analysis (e.g., Meylan 1987), suggests it should be the sole member of the genus *Trionyx* and considered an early offshoot of the group. This species is aquatic with a native range covering much of the African continent except the waterways of southern and northwestern Africa. The little known about the species' natural history is reviewed in Burghardt et al. (1996). In the 1980s keepers at the National Zoo observed Pigface raking his foreclaws into the flesh on his neck and also biting his forelimbs. Considerable damage to the skin and muscle was observed, along with infection and fungal growth. Similar sterotypical and self-injurious behavior has long been noted in captive birds and mammals including feather plucking and hair pulling (Hediger 1950) but is rarely reported in reptiles. In 1986, in an attempt to stop this detrimental behavior, Dale Marcellini, Reptile Curator, suggested that keepers add objects to the turtle's tank to see if Pigface could be distracted from his self-injurious behavior. The level of self-mutilation behavior, as shown by sores and wounds, did decrease. However, when the objects were continuously available for long periods, Pigface would interact less with them and the self-mutilation would increase. Thus, although the provisioning of objects continued on a fairly regular basis, it was interspersed with breaks of up to several weeks duration with no added objects. When I visited the Zoo and saw the animal interact with the objects I was immediately impressed with the similarity to the way aquatic mammals interacted with objects in water. As no documentation was available, we video-taped and analyzed extensive sequences of Pigface's behavior (Burghardt et al. 1996).

The video record included periods when several objects were added to the tank in addition to submerged sticks that were always present. These objects included a familiar standard sized brown basketball, an unfamiliar orange basketball of the same size, and a round hoop prepared from a 2 meter section of garden hose partially filled with water (so that it would float vertically as well as horizontally), the rubber fill hose, and fish. We analyzed the video, constructed an ethogram, and compiled an activity budget. A sequential record of the video was then constructed identifying where, when, and in what order the activities were performed. From these data, the duration and percentage of time the turtle spent performing each activity, the percentage of time spent interacting with an object when it was available, the frequencies and rates of approaches to the objects, and the frequencies and rates of contact with the objects were calculated.

An activity budget showed that Pigface interacted with the objects a great deal of the time. He was active over two-thirds of the time and he interacted with the objects 31% of the time (Fig. 1.1). The proportion of time spent with the different objects is even higher if calculated only for periods that the objects were available.

What did Pigface do with the objects? In addition to frequently approaching, following, and visually inspecting them he would nose, bite, grasp, chew, push, pull, or shake them with his mouth. He would also use his forelimbs to hold down, to draw closer, or to otherwise maintain contact with objects. The specifics varied with the object. Sticks were recipients of all the behaviors listed but basketballs were only nosed, pushed, and bit (Fig. 1.2). When the hoop was present it was highly favored. When floating vertically the turtle nosed, bit, chewed, shook, pushed, pulled, and sometimes repeatedly swam through it.

One of the criteria often proposed for play is that it is fun or pleasurable to the participants. Clearly this is hard to demonstrate in nonhuman species and especially in a species where it is difficult to anthropomorphize gestures and facial expressions. But the issue of the private experience of animals is returning as a meaningful one in nonhuman

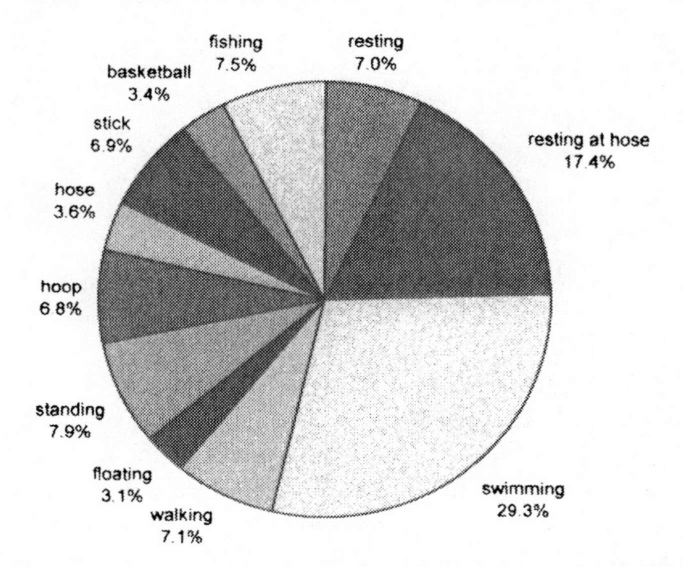

Fig. 1.1. The activity budget of a captive Nile soft-shelled turtle (from Burghardt et al. 1996).

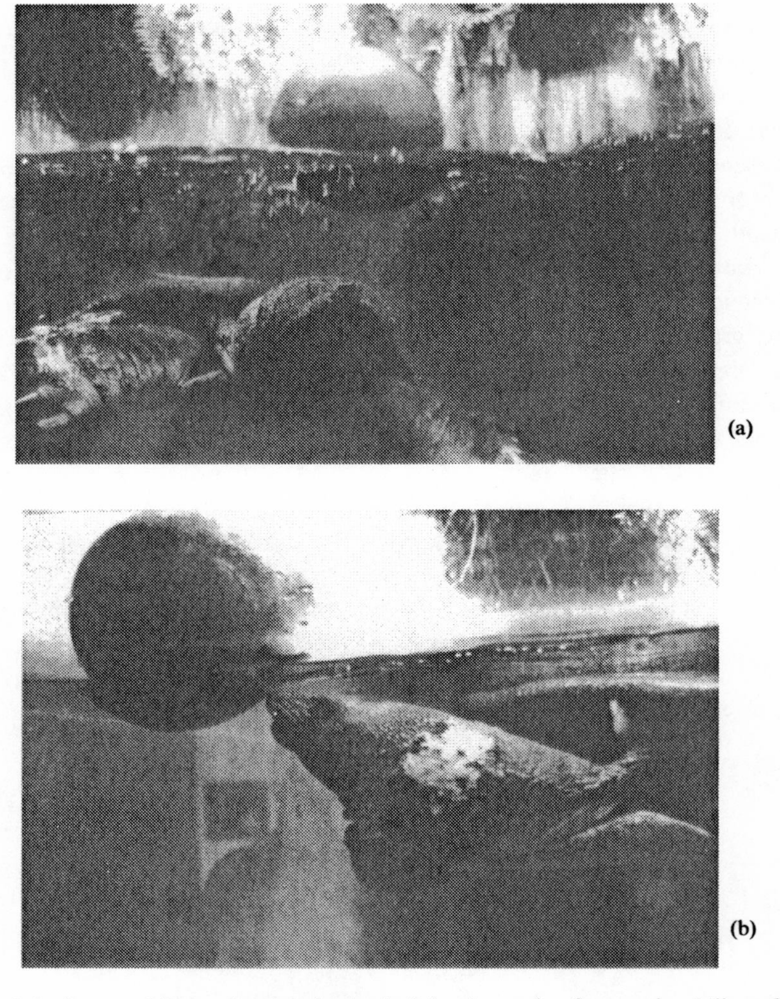

Fig. 1.2. Approaching a floating basketball in the tank of a captive Nile soft-shelled turtle (from Burghardt et al. 1996).

species (Burghardt 1997). Does Pigface show any 'joy?' Consider this. When the tank was being refilled with water, which streamed from a hose at the bottom of his tank, Pigface would orientate towards it so that the stream of water flowed over his head. He appeared unsatisfied until he had adjusted the direction of the hose so it was 'just right.' When everything was set he would remain there motionless for some time. This was rather striking given his usual high rate of activity. However, when the water was turned off, the turtle rapidly became restless and moved off. Could this be akin to our pleasure with the wind in our face on a spring afternoon? Your call. But this is a species often found in rivers where there are currents. His enclosure had none.

Regardless, by any traditional criteria the turtle 'played' with the various objects in his tank and, furthermore, this behavior made up a considerable portion of his activity budget. This is a large amount of time for any animal to spend in play behavior. Mammals, including primates, typically play 1–10% of the time (Fagen 1981). If play behavior was to occur in reptiles, Fagen (1981) suggested that it would be seen in crocodilians, because of their extensive parental care, and varanid lizards, because of their relatively high, for reptiles, metabolic rate. Thus the two examples Fagen cited fit SRT. What is interesting about these examples and Pigface is that none of these instances involved neonate or very young animals. While adult play in mammals, especially in captivity, is not uncommon, it is definitely most conspicuous as a phenomenon of the juvenile period, being transient in many species.

What factors facilitated the occurrence of this play behavior in Pigface? Several possibilities derive from those previously identified as factors in endothermic animal play (Burghardt 1988). These include its occurrence in an energy-efficient medium for locomotion (water), the more than satisfactory nutritional and thermal status in captivity, the spartan captive environment, and the prior evidence of stress (boredom?) as indicated by the earlier history of self mutilation behavior. One supportive example for the role of water in facilitating play is the greater amount of aquatic versus terrestrial play in aquatic mammals such as seals (Renouf 1993).

Since we published our report on Pigface several other reports have come to my attention. Although *Trionyx triunguis* is very infrequently kept in captivity (ISIS records for 1995 list less than 10 worldwide), two adult Nile soft-shelled turtles at the Toronto Zoo have been observed interacting with a floating ball and hoops similar to Pigface (R. Johnson, pers. comm.). I have viewed a videotape taken in Oct. 1996 that confirms the similarity. As Pigface's recent death has precluded further study with him, my laboratory is collaborating on a more experimental study of these animals at the Toronto Zoo.

Mann & Mellgren (in press) report that young loggerhead (*Carretta caretta*) and green sea turtles (*Chelonia mydas*) approached and contacted small objects in a manner reminiscent of object play as well as curiosity (c.f. Glickman & Sroges, 1966; Chiszar et al. 1976). They discuss whether or not the behavior with the objects is play or autogrooming, since the turtles made contact with the carapace and flippers by using rubbing motions. The authors systematically added and removed objects. The results showed that initially the turtles were neophobic, but as time went on they began to seek out and then to interact with the objects more and more frequently. Thus these observations tracing the initial reaction to objects and their later development separate out the distinction between investigation and play proposed by Hutt (1966) in her studies of children. That the behavior is not just an artifact of captivity is supported by professional scuba divers who have observed adult green turtles approaching and behaving similarly to objects in the ocean, and who also interpret the behavior as both play like and pleasurable (R. Mellgren, pers. comm.).

Locomotor play

The behavior of Pigface with the hoop and his frequent swimming, floating and other locomotion may be considered locomotor play akin to soaring in birds and leaping and brachiating in mammals. But locomotor/rotational play seems invoked as a label only when dramatic or exaggerated movements are involved, an often problematic task in little known species. More anecdotally, J. Loeven (pers. comm.) observed a captive wood turtle (*Clemmys insculpta*) that repeatedly climbed and slid down a board into the water. Interestingly, wood turtles have a reputation for being particularly clever among turtles. More generally, many aquatic turtles 'spontaneously' swim in a three dimensional space. Certainly Pigface spent much time swimming, although he should have known he was not going anywhere. Is this locomotor behavior comparable to the soaring of hawks, the brachiation of gibbons, and the gamboling of colts? All these have been considered play (Beach 1945). On the other hand, locomotion in fish, except for leaping out of the water, has not been considered playful. Even the latter has been questioned as merely parasite removal behavior (Fagen 1981). But the attribution of a plausible current function to a behavior as a means to deny the label play is not altogether satisfying, although very common.

Social play

Pigface, although raised apart from other turtles, did develop a relationship with his keeper. The rubber hose used to refill the tank in the morning session was lowered from the top of the tank with the outlet resting on the substrate. The turtle nosed, bit, clawed, and push/pulled the hose. If the keeper pulled the hose when it was in the turtle's mouth, almost pulling him out of the water, the turtle would respond by trying to swim backwards, pulling the hose back into the tank. This was reminiscent of a tug of war game. This took place fairly regularly according to the keeper.

As a possible example of conspecific social play in turtles, I will summarize what we know about the juvenile courtship-like behavior reported in some aquatic species of North American emydid turtles. Although a detailed report is being published elsewhere (Kramer & Burghardt, 1998), a summary is included here to stimulate discussion on its possible playful nature and to foster more such observations.

A number of emydid turtles engage in precocious courtship well before they attain sexual maturity and develop morphological secondary sexual characteristics. Genera reported include *Chrysemys*, *Graptemys*, *Pseudemys*, and *Trachemys*. Pellis (1993) reviewed precocious sexual behavior in muroid rodents and argued that in this group such behavior was the origin of juvenile social play. Could precocious courtship in turtles be productively viewed through the same lens?

First, it is useful to describe sexual dimorphism and adult courtship in these turtles. The dimorphism and courtship of adult *Pseudemys nelsoni* (Kramer & Fritz 1989) are typical for aquatic species of emydid turtles. Mature females are usually larger than males. Adult males have an enlarged tail and very elongated foreclaws. Both characters are related to courtship and copulation, and are not involved in intermale agonistic competition. Courtship begins with the male's approach and investigation of the female. This proceeds to active following, interspersed with the courtship display termed 'titillation'. During titillation the male swims above and parallel to the female, facing the same direction, and periodically and repeatedly thrusts his front limbs forward, rotating them so that the palms face out, and rapidly vibrates the digits of his front feet. The display is clearly oriented towards the head (eyes) of the female. The male's long front claws appear to emphasize the display. Titillation may continue for many minutes and may be repeated for hours. At some point the male may attempt to mount and, if the female permits it, to copulate. If unsuccessful, the male may continue to display or leave the female. Courtship of other species of *Psuedemys* appears to be similar. In *Chrysemys picta* and *Trachemys scripta* males face the females during titillation (Ernst 1971; Jackson & Davis 1972b) instead of swimming above them. Adult females always lack the elongated foreclaws and only very rarely have been observed titillating.

Observations of precocious courtship were made on four *P. nelsoni*, three *P. floridana penninsularis*, and two *P. concinna suwanniensis* at the University of Tennessee. The animals were observed for several months in an aquarium containing about 140 l of water. During observation sessions we recorded all interactions among turtles using paper and pencil or super 8 mm film. All occurrences and focal animal methods were employed.

Precocious sexual behavior occurred when one individual approached another and displayed (titillated) one or more times in a series of bouts. The vibration rate of the front claws of juvenile *P. nelsoni* (4.4–10.8 vibrations/second) are comparable across species and similar to adult male courtship rates. Both juvenile males and females approached and titillated other turtles. Longer duration bouts were found in juveniles than in the adult male observed. These preliminary observations suggest that juvenile titillation displays may be exaggerated compared to adult displays. Certainly they were not just tentative or brief intention movements or mere intimations of adult courtship.

Subjects displayed about once every 4–10 minutes in the two datasets. Since the choice of observation periods was based on times when social interactions were generally high, this does not adequately summarize precocious courtship frequencies throughout the day. But the behavior in *P. nelsoni* was neither rare or sporadic. Displays occurred in bouts lasting one minute or more. A display comprised 1–6 individual acts in each of which the front claws were vibrated several times.

The titillation display often terminated when the displaying animal abandoned the recipient. The response to the displays varied. The recipient often appeared to ignore the displaying individual, retracted or rubbed its head, or pushed away the displaying animal with a front leg. Less often, a recipient who appeared to be more disturbed by the display, turned or moved away from the displaying individual. In 18 of 213 displays the recipient gaped or snapped at an approaching or displaying individual. On one occasion, a recipient bit the displayer on one of its front feet. Gaping, snapping, or biting invariably halted further or incipient displays to the recipient, nor would this recipient be approached by the displaying animal again for several minutes. Mounts rarely followed displays; only two were seen during the entire period the juveniles were under observation and neither involved cloacal contact. In both instances, one of the female *P. nelsoni* briefly mounted the male *P. nelsoni* following several bouts of displaying.

Overall, the behavior was at least as vigorous as adult courtship and involved considerable maneuvering of the displaying turtle to maintain its relative position with respect to the often moving recipient. Precocious courtship appeared to be the most energetic behavior performed by these animals. Although the energetic cost of a display was not directly measured, the turtles surfaced for air significantly more frequently in the 30 second period following displays than they did during other 30 second periods.

The youngest displaying *P. nelsoni* we observed was wild-caught and about five months old. It displayed briefly to a piece of dried dog food in the water before biting it. Cagle (1955) observed a much younger (35 day old) *P. concinna suwanniensis* displaying to a snail before eating it. One of the laboratory reared *P. nelsoni* in this study, when about one years old, displayed to a live cricket before eating it. Petranka & Phillipi (1978) saw titillation to inanimate objects in *P. concinna* and *P. floridana*. D. Jackson (pers. comm. to M. Kramer) observed many instances of precocious courtship in young *P. nelsoni, P. alabamensis, P. concinna*, and *Trachemys scripta*, starting from three weeks of age. Rives (1978) also saw displays directed to inappropriate objects in smaller wild-caught turtles. Thus not only can the display be performed by very young turtles, the stimuli eliciting it may also become more specific as the turtles mature.

Precocious titillation was much more frequent in *P. nelsoni* juveniles than in the other two species. Additionally, *P. nelsoni* juveniles displayed more often to conspecifics than to heterospecifics (210 vs. 3 displays). Thus both species differences and conspecific recognition occur in juvenile courtship behavior. There were also partner preferences among the four juvenile *P. nelsoni*.

Although clearly identifiable as similar to adult courtship, precocious courtship behavior of juvenile turtles differs from adult courtship in several ways. The initiation phase (Kramer & Fritz, 1989) is cursory or absent, although individuals often sniffed the cloacal or head region briefly prior to, but rarely during, a display.

A most fascinating aspect of young–adult differences, besides the sex difference, was in how the turtles oriented while performing the titillation displays. In juvenile *P. nelsoni* and probably other species (Rives 1978), titillation displays occurred from two positions: (1) head-to-head: the displaying animal facing the recipient and (2) swim above: the displaying animal swimming above the recipient and facing in the same direction (Fig. 1.3). The first position is characteristic of adult *Chrysemys picta* (Taylor 1933), some *Trachemys* (Jackson & Davis 1972b) *Graptemys*

Fig. 1.3. The two positions adopted during precocious courtship in *Pseudemys nelsoni* (from Kramer & Burghardt, 1998).

(Vogt 1978); the second position is characteristic of all *Pseudemys* species in which courtship has been observed (Marchand 1944; Jackson & Davis 1972a; Kramer & Fritz 1989). Most filmed juvenile *P. nelsoni* displays were head-to-head; this is not the orientation used in adult courtship in this species. More comparative work is needed to determine if the occurrence in juveniles of precocious courtship postures, not found in adult conspecific courtship, is due to the retention and ritualization of plesiomorphic behavior patterns. Courtship in adult and juvenile turtles could be an important source of evolutionary information about the phylogeny and ritualization of courtship displays. Perhaps precocial courtship retains ancestral traits lost in the more stereotyped 'serious' adult behavior.

There were other differences from adult courtship. Juvenile *P. nelsoni* displayed only to turtles on the substrate, whether moving or stationary. Males in the field and laboratory displayed only to swimming females (Kramer & Fritz 1989). On occasions, the interaction between two juveniles attracted a third, who started displaying to one of the other two. On one occasion all four juveniles were clustered together, with at least three displaying. Adults were not seen to join courting pairs in the field.

In spite of the differences from adult courtship, the precocious sexual behavior of *P. nelsoni* shares most of its components with adult courtship and not with other adult or juvenile behaviors. In particular, precocious sexual behavior resembles the swim above phase (Kramer & Fritz 1989) of adult courtship. The sole recorded instance of an adult female *P. nelsoni* displaying (Jackson 1977) was more similar to the juveniles' displays than those of adult males because this female also displayed from the head-to-head position.

One interpretation of precocious courtship was put forth by Rives (1978). He experimentally paired unfamiliar wild-caught juvenile *Chrysemys picta* and *Trachemys scripta* and recorded displays. Rives' observations agree with ours in documenting that males and females displayed at about the same rates and that conspecifics were the preferred recipients. Based on more qualitative observations of long-term captive turtles, Rives suggested that titillation was aversive to recipients and might serve to establish and maintain social dominance, indicating that this courtship behavior was precociously borrowed for a different social function. He found that large animals displayed to smaller ones. In contrast, our observations were on *P. nelsoni* with juveniles with a long history of prior social interactions with one another. While we also found recipients sometimes avoiding or reacting against the displayer, no dominance order or size relationship was seen. However, recipients of displays in all these species often responded negatively to the displayer. Thus the displayer, but not the recipient, seemed motivated to engage in precocious courtship behavior and found it in some way rewarding, if not exactly pleasurable.

There are virtually no observations in the field of the underwater behavior of juvenile aquatic emydid turtles and thus the possibility that the displays were artifacts of captivity cannot be eliminated. However, precocious courtship has been observed in wild-caught turtles as well as captive-hatched and reared animals in several species. For example, Rives (1978) found the behavior readily in his extensive experiments and observations on field-fresh animals of several species. This being said, field observations are needed if, as argued below, this behavior may have some developmental role.

Does precocious courtship in turtles share characteristics with play? Pellis (1993) compared true play, as found in rats, which is elaborate, ritualized, and prominent in the behavioral repertoire, from the brief, sporadic, and simple pre-copulatory behavior seen in juvenile voles with intermediate stages found in hamsters and gerbils. From his careful structural descriptions, he argued that many elements of play, even play fighting, have their origin in sexual rather than agonistic behavior. Precocious sexual behavioral patterns already present in immature animals were thus viewed as the bases on which sexual play and play fighting were derived. According to this scenario, in the later evolutionary stages such juvenile social play became a necessary developmental precursor to successful sexual behavior as an adult and could be valuable in other social contexts as well (see also Pellis & Pellis, Chapter 6). He pointed out that simple precocious sexual behavior was uncommon, sporadic, and brief whereas 'true' play was frequent and time-consuming.

If the turtles were rodents, where would precocious courtship behavior by turtles fit in Pellis' (1993) scheme for the evolution of play fighting? Precocious courtship behavior occurred frequently in *P. nelsoni* and differed in obvious ways from adult courtship behavior. This suggests that it might be characterized as part of rudimentary or true play using Pellis' scheme. However, the fact that mutual titillation was seen by Rives (1978) in other genera indicates that the 'social' nature of this social play needs to be evaluated more closely.

Precocious courtship appears limited to sexually immature animals and is virtually never performed by adult females. Thus unlike the object and locomotor play described above, this behavior seems clearly limited to early ontogeny. If the behavior is playful, what is its function? Elsewhere various functions such as maturation, practice, and motor training are discussed in some detail (Kramer & Burghardt 1998). Here only the motor training hypothesis as recently updated by Byers & Walker (1995) and Byers (Chapter 10) will be discussed. These authors presented evidence suggesting that play behavior in some juvenile mammals may permanently modify muscle fiber differentiation and cerebellar synapse distribution. The evidence is limited to the correlation of peak play periods with the onset and completion of these physiological processes. Unlike most motor/neural/physiological training functions of play that have been advanced, however, the changes posited by Byers & Walker (1995) are permanent and unlikely to be induced by other behavioral means. Could precocious courtship in turtles serve some similar function? Perhaps precocious courtship behavior is a means of consolidating, not specific behavioral routines (the titillation display is in adult form when first performed), but the neural (cerebellar) and muscular substrates necessary for complex negotiation in a three dimensional world. If so, then this explanation might account for display behavior in juvenile females. Of course, one can posit other functions that explain female performance. Perhaps juvenile females are role-playing; performing male typical routines may help them in courtship and mate choice when they are adults. This seems rather improbable, perhaps, but at this stage of our ignorance all possibilities should be brought out.

Although the motor performance of the titillation display is adult-like when first observed, the orientation and movements associated with this behavior may need to be practiced. The fact that juveniles titillated animals on the substrate rather than while swimming suggests that they may need to perfect performing the behavior while actively negotiating and maintaining contact with a swimming animal. It may be similar to the childhood game of circling one hand on one's chest while using the other hand to pat the top of one's head with an up-down motion. Either one is easily performed singly; doing both simultaneously needs practice. There is evidence that the stimulus control for precocial courtship shifts with age. The youngest animals display to many small objects while older juveniles and adults display mostly to conspecifics. The high frequency of this behavior among juvenile *P. nelsoni* allows each individual to have many interactions with others, providing substantial opportunities for learning to occur. This explanation, however, fails to account for display behavior in juvenile females, nor does it explain partner preferences. Although observations of adult courtship are clearly deficient, there is no evidence whatsoever to suggest that this display is used in contexts other than courtship in nature.

The evolution of play

That behavior with mammal-like overtones is found in turtles may not be surprising since, with the exception of parental care and high metabolic rates, they share some traits with the more playful species in the mammalian radiation. The social system of *P. nelsoni* (and perhaps other species) appears complex (Kramer 1986) and may be mediated through individual recognition (Kramer 1989). Turtles are considered to be 'intelligent', suggested by the consistent bias of comparative psychologists to use them rather than other reptiles (Burghardt 1977). Of all extant reptiles, the turtle brain is probably the 'closest' to that of mammals (Butler & Hodos 1996) and there are homologues of all major brain structures. Aquatic turtles live in an energy efficient medium and are less subject to the need for adjusting to rapid temperature fluctuations confronting terrestrial ectotherms. Thus turtles live in an environment where some of the metabolic advantages of endothermy are less critical for life style differentiation. Aquatic turtles have evolved physiological adaptations that provide for greater aerobic metabolic rates and thermoregulation than found elsewhere in nonavian reptiles (Gatten 1974; Smith et al. 1981; Stone et al. 1992). Finally, turtles are both long lived and take many years to reach sexual maturity, which would give any of the proposed long range benefits of play (see Fagen 1981) sufficient time to accrue. Many of these features are also found in crocodilians.

A summary of the features which are similar between typical play criteria and the two main types of putative turtle play described here are presented in Table 1.1. These criteria are taken from Burghardt (1984) and, while not exhaustive, do provide an unbiased list on which to map both object and social play. Kramer & Burghardt (1998) carry out a similar exercise with a list derived from Fagen (1981) that has more features specific to social play. The results are similar: these examples of turtle behavior fulfill the requirements sufficiently to be considered play.

Pigface's negative reaction to the lack of stimulation and the ameliorative value of providing objects suggests that some reptiles may have innate needs for stimulation and activity that, if not expressed, could be

Table 1.1. *Characteristics of play extracted from the literature (Burghardt 1984) and applied to two kinds of play like behavior in turtles. X – applies; ? – may apply; NA – not applicable; O – not observed.*

Attributes of Play	Soft-Shell Object Play	Precocious Courtship
No obvious immediate function	X	X
Sequentially variable	X	X
Quick and energetically expensive	X	X
Exaggerated, incomplete, or awkward	X	X
Most prevalent in juveniles	O	X
Breakdown in role relationships	NA	X
Special 'play' signals	O	O
Mixing of behavior patterns from several contexts	X	X
Relative absence of threat and submission	X	?
Relative absence of final consummatory acts	X	X
Stimulation seeking	X	X
Pleasurable affect	?	?

detrimental. These may differ quantitatively and qualitatively between active foragers and more sit-and-wait (ambush) predators. In both the object and social contexts the turtles performed behaviors that were normally used in other contexts such as exploring, chasing, attacking, biting, dismembering, orienting, and titillating. These contexts provide energetically and metabolically fit animals the opportunity to perform behaviors that are part of their normal repertoire and motivational system. The lack of appropriate stimulation may block their expression and some types of stimulation may encourage their expression. Detailed work needs to be done but this is just beginning with nonavian reptiles (Chiszar et al. 1993). More observations of behavior in both young and adult turtles should be sensitive to the comparable phenomena studied as 'play' in birds and mammals.

At the symposium one of the editors asked whether play should be considered a plesiomorphic trait found in the common ancestor of turtles and mammals. Turtles are the descendants of the oldest extant reptilian lineage (Anapsida) closest to the stem group (Synapsida) leading to the therapsids and modern mammals (Gautier et al., 1989). The Testudines lineage may go back to the late Triassic, over 200 million years BP (Rougier et al., 1995) and the Mammalia to at least 200 million years BP. This common ancestor of turtles and mammals probably lived in the Permian era, over 300 mya.

In spite of the ancient bauplan of these turtles and their ancestral closeness to the common ancestor of mammals and reptiles, I do not think that play is a plesiomorphic character for amniotes. First, there is no evidence as yet for playfulness in amphibia. It is also doubtful whether playfulness is synapomorphic for amniotes since there is little evidence from other turtle species, other reptiles (including primitive birds such as ratites) and primitive mammals (e.g. monotremes). The fact that play has only been reported in a few reptiles and documented in less than a handful suggests that it has evolved independently several times. It is more conservative to view play in isolated taxa as independently derived when a series of ecological, life history, and physiological factors coincide. But such isolated situations are nonetheless important, for they can allow us to explore the extent of the similarity in underlying mechanisms and in the developmental precursors and functional consequences of such playfulness. Thus the various kinds of playfulness should be considered neither a 'legacy from the past' nor an 'adaptation to prepare for the future.' Different lineages may share both factors but in varying degrees. What the observations on turtles suggest is that the primary processes shaping play run deeper in evolutionary time than has been typically appreciated and that the difficulty of demonstrating clear adaptive functions for play may reside in the atavistic nature of some types of play in some animals. Its maintenance in a species may be a product of both history and lack of selection against it rather than strong selection for it, at least in some animals.

But there is a much more pressing issue that must be clarified before a phyletic approach to play, even among mammalian groups such as rodents, ungulates, or primates, can hope to be successful. Play is almost certainly a heterogeneous category linked together by characteristics that may be superficially similar, but have separate origins, causes, functions, and ontogenies. Play can be expressed in bodily activity (locomotor / rotational play), with objects that often are, can, or might be salient in the animals' lives (object play) or interactions of a social nature (social play). Whether these play like characters are contingent on a vertebrate style brain is not known. What I think the evidence presented here shows is that play like phenomena do occur in turtles, the extant survivors of the most ancient reptile lineage for which living representatives exist. The archaic 'anapsid brain' may not only have had the potential for rudimentary play but, insofar as this potential was shared by the reptilian ancestors of mammals, might itself have contributed to the evolution of quantitative advances in cognitive abilities eventually to be found in the mammals of subsequent eras. Play in crocodillians could also prove to be more common than thought and share derived features with birds. This raises the issue of how comparable are the 'play like' responses of turtles, crocodilians, and even lizards, whose common ancestors are similarly remote.

Progress in understanding the evolution of play will remain limited until the three traditional categories of play are deconstructed and the nature of similarities among them evaluated at all levels from physiology to ecology. This means that careful description of the ontogeny and topography of specific details of play need to be made available so that a truly phylogenetic analysis can be carried out. Furthermore, such details of playfulness must be correlated with other relevant life history, ecological, physiological, and psychological attributes. A good start can be made by collating the available data, looking for the most glaring lacunae, and then going out and filling them. As we do this, the role of reptiles would seem to be a necessary component of a successful analysis.

Acknowledgments

This chapter has benefited from conversations and comments of many individuals. I am particularly indebted to Matthew Kramer and Paul Weldon for their help. The preparation of this chapter and the research on which it is based were supported by grants from the National Science Foundation and the University of Tennessee Science Alliance.

References

Beach, F. L. 1945. Current concepts of play in animals. *Amer. Nat.* **79**, 523–541.

Bekoff, M. & Byers, J. A. 1981. A critical reanalysis of the ontogeny and phylogeny of mammalian social and locomotor play: an ethological hornet's nest. In *Behavioral Development. The Bielefeld Interdisciplinary Project* (ed. K. Immelmann, G. W. Barlow, L. Petrinivich & M. Main), pp. 296–337. Cambridge: Cambridge University Press.

Belew, R. K. & Mitchell, M. (eds.). 1996. *Adaptive Individuals in Evolving Populations: Models and Algorithms*. Santa Fe Institute Studies in the Sciences of Complexity, Vol. 26. Reading, MA: Addison Wesley.

Bennett, A. F. 1982. The energetics of reptilian activity. In *Biology of the Reptilia*. Vol. 13. (ed. C. Gans & F. H. Pough), pp. 155–199. London: Academic Press.

Burghardt, G. M. 1977. Learning processes in reptiles. In *Biology of the Reptilia. Vol. 7. Ecology and Behavior*. (ed. C. Gans & D. Tinkle), pp. 555–681. New York: Academic Press.

Burghardt, G. M. 1982. Comparison matters: curiosity, bears, surplus energy, and why reptiles do not play. *Behav. Brain Sci.*, **5**, 159–60.

Burghardt, G. M. 1984. On the origins of play. In *Play in Animals and Humans*. (ed. P. K. Smith), pp. 1–41. London: Basil Blackwell.

Burghardt, G. M. 1988. Precocity, play, and the ectotherm-endotherm transition: Profound reorganization or superficial adaptation? In *Handbook of Behavioral Neurobiology, Vol. 9. Developmental Psychobiology and Behavioral Ecology* (ed. E. M. Blass), pp. 107–148. New York: Plenum Press.

Burghardt, G. M. 1997. Amending Tinbergen: A fifth aim for ethology. In: *Anthropomorphism, Anecdotes, and Animals* (ed. R. W. Mitchell, N. S. Thompson, & H. L. Miles), pp. 254–276. Albany, NY: SUNY Press.

Burghardt, G. M. 1998. Play. In *Comparative Psychology: A Handbook*, (ed. Greenberg, G. & Haraway, M.), pp. 757–67. New York: Garland.

Burghardt, G. M., Ward, B., & Rosscoe, R. 1996. Problem of Reptile play: Environmental enrichment and play behavior in a captive Nile soft-shelled turtle. *Zoo Biology*, **15**, 223–38.

Butler, A. B. & Hodos, W. 1996. *Comparative Vertebrate Neuroanatomy: Evolution and Adaptation*. New York: Wiley-Liss.

Byers, J. A. & Walker, C. 1995. Refining the motor training hypothesis for the evolution of play. *Amer. Nat.*, **146**, 25–40.

Cagle, F. R. 1955. Courtship behavior in juvenile turtles. *Copeia*, 1955, 307.

Chiszar, D., Carter, T., Knight, L., Simonson, L. & Taylor, S. 1976. Investigatory behavior in the plains garter snake (*Thamnophis sirtalis*) and several additional species. *Anim. Learn. Behav*, **4**, 273–8.

Chiszar, D., Smith, H. M. & Radcliffe, C. W. 1993. Zoo and laboratory experiments on the behavior of snakes: Assessments of competence in captive-raised animals. *Amer. Zool.*, **33**, 109–16.

Ditmars, R. L. 1933. *Reptiles of the World: The Crocodilians, Lizards, Snakes, Turtles and Tortoises of the Eastern and Western Hemispheres*. New York: Macmillan.

Ernst, C. H. & Barbour, R. W. 1989. *Turtles of the World*. Washington, D.C.: Smithsonian Institution Press.

Ernst, C. H. 1971. Observations of the painted turtle, *Chrysemys picta*. *J. Herpetol.*, **5**, 151–60.

Fagen, R. 1981. *Animal Play Behavior*. New York, Oxford University Press.

Fagen, R. 1996. Animal play, games of angels, biology, and Brian. In: *The Future of Play Theory* (ed. A. D. Pellegrini), pp. 23–44. Albany, NY: SUNY Press.

Ficken, M. S. 1977. Avian play. *Auk*, **94**, 573–82.

Gatten, R. E., Jr. 1974. Effects of temperature and activity on aerobic and anaerobic metabolism and heart rate in the turtles *Pseudemys scripta* and *Terrapene ornata*. *Comp. Biochem. Physiol.*, **48A**, 619–48.

Gautier, J., Cannatella, D., De Queiroz, K., Kluge, A. G. & Rowe, T. 1989. Tetrapod phylogeny. In *The Hierarchy of Life* (ed. B. Fernholm, K. Bremer, & H. Jörnvall), pp. 337–51, New York: Elsevier.

Glickman, S. E. & Sroges, R. W. 1966. Curiosity in zoo animals. *Behaviour*, **26**, 151–88.

Groos, K. 1898. *The Play of Animals*. New York: D. Appleton.

Hall, G. S. 1904. *Adolescence, Its Psychology and Its Relation to Physiology, Anthropology, Sex, Crime, Religion, and Education. Vol. 1*. New York: D. Appleton.

Hatfield, J. W. III. 1996. *Green iguana – The Ultimate Owner's Manual*. Portland, OR: Dunthorpe Press.

Hediger, H. 1950. *Wild Animals in Captivity*. London: Butterworth.

Herzog, H. A. Jr. & Burghardt, G. M. 1977. Vocal communication signals in juvenile crocodilians. *Z. Tierpsychol.*, **44**, 294–304.

Hill, C. 1946. Playtime at the zoo. *Zoo Life*, **1**, 24–6.

Hutt, C. 1966. Exploration and play in children. In *Play, Exploration and Territory in Mammals* (ed. P. A. Jewell & C. Loizos), pp. 61–81. London: Symposium Zoological Society.

Jackson, C. G. 1977. Courtship observations on *Chrysemys nelsoni* (Reptilia, Testudines, Testudinidae). *J. Herpetol.*, **11**, 221–2.

Jackson, C. G., Jr. & Davis, J. D. 1972a. Courtship display behavior of *Chrysemys concinna suwanniensis*. *Copeia*, 1972, 385–7.

Jackson, C. G., Jr. & Davis, J. D. 1972b. A quantitative study of the courtship display of the red-eared turtle, *Chrysemys scripta elegans* (Wied). *Herpetologica*, **28**, 58–64.

Kramer, M. 1986. Field studies on a freshwater Florida turtle, *Pseudemys nelsoni*. In *Behavioral Ecology and Population Biology* (ed. L. C. Drickamer), pp. 29–34. Privat, I.E.C. Toulouse.

Kramer, M. 1989. Individual discrimination of in juveniles of the turtles, *Pseudemys nelsoni* and *Pseudemys floridana* (Chelonia, Emydidae). *Biol. Behav*, **14**, 148–56.

Kramer M. & Burghardt. G. M. (1998). Precocious courtship and play in emydid turtles. *Ethology*, 104, 38–56.

Kramer, M. & Fritz, U. 1989. Courtship behavior of the turtle, *Pseudemys nelsoni. J. Herpetol,* **23**, 84–6.

Lazell, J. D., Jr. & Spitzer, N. C. 1977. Apparent play in an American alligator. *Copeia* 1977, 188.

Lovich, J. E., Garstka, W. R., & Cooper, W. R., Jr. 1990. Female participation in courtship behavior of the turtle *Trachemys s. scripta. J. Herpetol.,* **24**, 422–4.

Mann, M. A. & Mellgren, R. L. in press. Sea turtle interactions with inanimate objects: Autogrooming or play behavior. In: *Proceedings of the 16th Annual Workshop on Sea Turtle Biological Conservation* (ed. R. Byles & C. Coogan), NOAA Technical Memorandum. Washington, DC: NMFS-SEFSC.

Marchand, L. J. 1944. Notes on the courtship of a Florida terrapin. *Copeia,* 1944, 191–2.

Martin P. & Caro, T. M. 1985. On the functions of play and its role in behavioral development. *Advances in the Study of Behavior*, **15**, 59–103.

Martins, E. (ed.) 1996. *Phylogenies and the Comparative Study of Behavior.* New York: Oxford.

Meylan, P. A. 1987. The phylogenetic relationships of soft-shelled turtles (Family Trionychidae). *Bull. Amer. Mus. Natur. Hist.,* **186**(1), 1–101.

Mitchell, R. W. & Thompson, N. S. 1991. Projects, routines and enticements in dog-human play. *Pers. Ethol.*, **9**, 189–216.

Ortega, J. C. & Bekoff, M. 1987. Avian play: comparative evolutionary and developmental trends. *Auk,* **104**, 338–41.

Pagel, M. 1994. Detecting correlated evolution on phylogenies: a general method for the comparative analysis of discrete characters. *Proc. R. Soc. Lond., B* **255**, 37–45.

Pellis, S. M. 1993. Sex and the evolution of play fighting: A review and model based on the behavior of muroid rodents. *Play Theory and Research,* **1**, 55–75.

Petranka, J. W. & Phillippi, A. 1978. Observations on the courtship behavior of juvenile *Chrysemys concinna concinna* and *Chrysemys floridana hoyi* (Reptilia, Testudines, Emydidae). *J. Herpetol.,* **12**, 417–9.

Renouf, D. 1993. Play in a captive breeding colony of harbour seals (*Phoca vitulina*): constrained by time or energy? *J. Zool. (London),* **231**, 351–63.

Riddle, D. L., Blumenthal, T. & Meyer, B. J. (eds.). 1996. *C. elegans II. Monograph* 33. Plainview, NY: Cold Spring Harbor Laboratory Press,.

Rives, J. D., 1978. A comparative study of courtship related behavior in immature emydid turtles of two species. Unpubl. M.Sc. Thesis. Lafayette, LA: University of Southwestern Louisiana, 115 pp.

Roggenbuck, M. E. & Jenssen, T. A. 1986. The ontogeny of display behaviour in *Sceloporus undulatus* (Sauria: Iguanidae). *Ethology*, **71**, 153–65.

Rougier, G. W., Fuente, M. S. de la & Arcucci, A. B. 1995. Late Triassic turtles from South America. *Science,* **268**, 855–8.

Schmidt-Nielsen, K. 1972. Locomotion: Energy cost of swimming, flying, and running. *Science,* **177**, 222–8.

Schusterman, R. J. & Gisiner, R. C. 1997. Pinnipeds, porpoises, and parsimony: animal language research viewed from a bottom-up perspective. In: *Anthropomorphism, Anecdotes, and Animals* (ed. R. W.

Mitchell, N. S. Thompson, & H. L. Miles), pp. 370–82. Albany, NY: SUNY Press.

Smith, E. N., Robertson, S. L. & Adams, S. R. 1981. Thermoregulation of the Spiny Soft-shelled Turtle *Trionyx spinifer*. *Physiol. Zool*, **54**, 74–80.

Smith, P. K. 1996. Play, ethology, and education: A personal account. In: *The Future of Play Theory* (ed. A. D. Pellegrini), pp. 3–21. Albany, NY: SUNY Press.

Spencer, H. 1872. *Principles of Psychology* Vol. 2, Pt. 2. New York: D. Appleton.

Stone, P. A., Dobie, J. L. & Henry, R. P. 1992. The effect of aquatic oxygen levels on diving and ventilatory behavior in soft-shelled *Trionyx spiniferus*, stinkpot *Sternotherus odoratus*, and mud turtles *Kinosternon subrubrum*. *Physiological Zoology*, **65**, 331–45.

Taylor, E. H. 1933. Observations on the courtship of turtles. *Univ. of Kansas Sci. Bull.*, **21**, 269–71.

Tinbergen, N. 1963. On aims and methods of ethology. *Z. Tierpsychol.*, **20**, 410–33.

Vogt, R. C. 1978. Systematics and Ecology of the False Map Turtle Complex *Graptemys pseudogeographica*. Unpublished Ph.D. dissertation. 375 pp. Madison: University of Wisconsin.

2

Play in common ravens (*Corvus corax*)

BERND HEINRICH and RACHEL SMOLKER

Biology Department, Marsh Life Sciences Building, University of Vermont, Burlington, VT 05405-0086 USA

Among birds, the most common types of play (locomotor, object and social) are found in those orders (Psittaciformes and Passeriformes) with the most developed forebrains (Ortega & Bekoff 1987). Among the Passeriformes, the corvids are considered to have the most complex play behavior (Ficken 1977). The raven, *Corvus corax* is the largest passerine and probably has the largest brain volume of any corvid. It also inhabits the greatest geographical range and the most diverse habitats. The raven may therefore be expected to show tremendous behavioral flexibility, perhaps acquired in part through play (Gwinner 1966, Ficken 1977, Ortega & Bekoff 1987). Given these considerations, ravens provide a particularly useful 'outgroup' for comparison with mammals, as well as other birds.

In putting together this review of raven play behavior however, we were faced immediately with the problem of defining 'play'. 'Play' is notoriously difficult to define (Fagen 1981, Bekoff & Byers 1981, Bekoff 1984, Martin & Caro 1985; see also Bekoff & Allen, Chapter 5). We all recognize it at the extremes, but cannot define it clearly enough to fit it into an exclusive and objectively defined category of behavior. Ficken (1977) points out, and we agree, that play is even more difficult to identify in birds than in mammalian species. Perhaps the most widely accepted definition of play is as follows: '. . .all motor activity performed postnatally that appears purposeless, in which motor patterns from other contexts may often be used in modified forms or altered sequencing' (Bekoff 1984). This definition seems problematic in that it is not clear what is meant by 'appears purposeless,' and is therefore very difficult to apply. Behaviors may appear purposeless for various reasons including (a) the observer simply has not figured out what the purpose (benefit) is, (b) the benefit is not immediate (as when an animal practices and refines a

Fig. 2.1. Play in the nest includes object manipulation, vocal monologues and wing flapping.

skill that will lead to benefits at a later time), or (c) the benefits are multiple and confounding, as in play fighting where animals not only hone skills for later use in serious fighting, but may simultaneously be establishing important social relationships with peers as well as building muscle tone and coordination needed to escape predators.

Is play purposeless? Martin & Caro (1985) and Caro (1995) suggest that there are only minor costs in time and energy involved in play, and that play therefore may be maintained even when only minor benefits are accrued. Bekoff (1984) does not claim that play is purposeless, but rather that it appears purposeless. The animal *seems* to engage in the behavior 'for the fun of it'.

If we could come up with a workable definition of fun and measure it objectively, we would still be left with the begging question 'why is this *particular* behavior 'fun'?' If something feels good, an animal is likely to engage in the behavior more (take sexual intercourse, for example). Presumably, engaging in the behavior more frequently will only be favored if, in fact, the behavior has a benefit in terms of ultimate reproductive success. The fact that play seems to be 'fun', and therefore is engaged in at some costs (not necessarily minor), implies that it is very much a 'purposeful' activity, at least in terms of ultimate function.

The problem with the above definition (as with all other definitions of play we know of), is that we are left with no objective and applicable criteria for determining what is and what is not play. Our choice of what behaviors to include in this review of raven play is therefore necessarily subjective. The behaviors we choose here are those that strike us as obviously 'playful', and we believe that most observers would also immediately recognize them as play, even if reasons for doing so might be obscure. Nonetheless, we hope that readers will be left with some sense of raven play that will be useful for comparisons with other species. We discuss possible functions of each different type of play behavior in terms of 'serious'/adult raven ecology and behavior. These behaviors are illustrated in the figures.

Raven play behaviors

We begin with two typical notebook entries, from observations of ravens held in a large outdoor enclosure, to provide the reader with a sense of the nature of raven play.

14 December, 1993. (Four ravens, 7-months post-fledgling are involved here). Lefty held a piece of bark in her left foot, and dangled it down below the perch as she stood on her right foot. She then pitched over forward to hang upside-down with one foot, leaving the other foot with the bark free in the air. Then she let go, dropping to the ground, picking up and alternately carrying and stepping on the same piece of bark. She pecked the bark, carefully excavated snow from under it, then picked it up and rolled onto her side while grasping it with both feet. She finally left the bark and picked up a stick instead. Fuzz then took the bark but Houdi tried to claim it, and a tug-of-war ensued. Houdi won and took the bark up to a perch, but dropped it. Fuzz pounced on it, then lay on his side in the snow, holding it with both feet and manipulating it with his bill. Houdi took it back, and flew off with it in her bill. Lefty then chased Houdi all over the aviary. But Goliath snatched it away from Houdi, and then rolled onto his side,

Fig. 2.2. Object play in a variety of contexts. (a) Caching of inedible objects. (b) dropping and catching objects in flight. (c) tug of war. (d) rolling over on back with stick.

holding and pecking it. The three others ignored him, however, feeding on a dead calf. Goliath continued to manipulate the stick alone, eventually flying off the ground onto a perch where he held it in his dangling right foot. None of the others responded. He then flew back down to the ground, and jammed the stick into the snow out of sight. Within a half minute, he returned to dig it back out with back-and-forth sideways swipes of his bill. After recovering it, he hid it in a new place, covered it with other sticks, and then finally joined the others to feed on the calf. As soon as he did so, Fuzz left the calf, walked to Goliath's cache, removed the conspicuous sticks on top of the snow 'hiding' the cached stick, and recovered it. Houdi accompanied Fuzz, but Fuzz got possession of the stick. He dropped it while manipulating it, and Houdi instantly grabbed it. Fuzz picked up another stick, and Houdi dropped hers to investigate the new stick that Fuzz had. This new stick was not 7 cm long like the first – this one was nearly 1m long. Houdi and Fuzz each took hold of opposite ends, and pulled backward. The tug of war was won by neither. Instead, Goliath returned and took it away from both of them. They almost immediately gave it up without struggle and went to feed instead. Goliath rolled onto his back holding the new long stick, then dragged it all over the aviary while the other three fed, but soon dropped it and flew over to join the other three birds feeding at the calf. He perched on top of the calf while the other three fed from around it. Then Goliath flew off again to manipulate and carry yet another stick. All of this had taken just ten minutes. They often continued like this for hours, usually to the accompaniment of a variety of low soft vocalizations.

1 May, 1995. It is comical to watch Houdi trying to hold two sticks simultaneously. (During nest building the birds often carry 2–4 sticks at the same time to the nest.) Goliath often tries to take what she has. On one occasion, they had a tug-of-war with the sticks. Goliath won, but in the process, dropped one of the two sticks, which she retrieved. He then dropped the one he had, and tried to get the one she now had. She let him have it, but quickly picked up the one he dropped. Houdi then examined a 120 cm piece of PVC tubing, picked up a tennis ball from the other end of the aviary and shoved it (a tight fit) into the tube, adding dead dry leaves behind it. She then pulled first the leaves and then the ball back out of the pipe. Then she replaced the ball, removed it again – and again! Fuzz then came in from the adjoining aviary, and *he* pulled the snug-fitting ball out, and then put it back pounding it with his bill and adding debris behind it. He pulled it back out, put all back in, took all back out, and put it back in one more time. Houdi came and inspected. Fuzz then inspected also, put more debris and leaves behind the ball, pulled it all out, pounded the ball back in, took it out, and then put it into the other side of the PVC tube.

When I checked in the afternoon, the ball was in the tube and I knocked it out. They all ignored it. When I later put wood frogs (*Rana sylvatica*) into the tube, one at a time, they had no trouble catching them, after running from one end of the tube to the other, inspecting it from both ends. Through play, they had learned not to fear the tubing (they were very shy of it for the first several days), and now used it as an aid in prey capture in a unique circumstance.

Both of these accounts involve behaviors typical of courtship, food caching, dominance interactions, nest building and foraging activities. In these cases however, the behaviors are intermingled, out of context, and do not lead immediately to the benefits they are likely associated with in adults. Because the ravens are captive and therefore relieved of survival pressures such as food finding, the amount of time spent at play may be greater than in wild birds (Barrett et al. 1992).

We describe below several broad categories of behaviors that can be considered play or that contain components of play, and illustrate these with specific anecdotal observations.

Object exploration and manipulation

Some individual young ravens prior to fledging already use their bills to manipulate the sticks comprising their nest. They may spend several minutes at a time mildly pecking at and pulling on twigs, even just after being fed to repletion. After leaving their nests, many of their waking hours are spent in object manipulation. Of 10 young captive ravens living in a natural enclosure (but sometimes seeded with unnatural objects), all manipulated almost every kind of object they encountered, including leaves, pebbles, twigs, bottle caps, glass chips, pebbles, sea shells, dried beans and inedible berries (Heinrich 1995) during their first week after fledging. They exhibit great attraction to anything 'new' and rapid loss of interest in inedible objects that they are already familiar with. Some objects are cached, much as both adult and young birds do with excess food. Caching may involve shoving the item into a crevice, a hole in the ground, under leaves or snow, and covering it with debris.

The young birds seemingly obsessive drive to contact and manipulate literally all kinds of objects (conspicuous or cryptic, edible or inedible) has been described as 'neophilia' (Heinrich 1995) and also as 'play' in Australian magpies, *Gymnorhina tibicen* (Pellis 1981a,b). This object play is not motivated directly by hunger, occuring ubiquitously in well-fed birds (Heinrich 1995). Ravens continue to contact and manipulate objects as they get older, but their tendency to do so is increasingly tempered by caution when dealing with objects they have not experienced as young birds. After their first year or so, the ravens we studied in Maine were highly fearful of objects that they did not have contact with during the first few months out of the nest (Heinrich 1988, Heinrich et al. 1996).

Ravens are widely distributed throughout Eurasia, North America, and North Africa. They live in the polar regions, in the high elevations of the Himalayas, in deserts, forests, tundra, sea coasts, and in suburban settings in close association with humans. In addition to all of this geographical and ecological diversity, most ravens also experience a great deal of seasonal diversity (they do not migrate). As a result, potential food sources and dangers are not static. In experiments (Heinrich 1995), young forest ravens were presented with a variety of novel objects from a sea coast environment. At first they attended preferentially to all novel objects, but within minutes concentrated their attention on the one edible object in the collection. These experiments illustrate that object play and manipulation serves to familiarize young birds with potential food items that may be unique to their ecological circumstances.

Play caching

Young ravens routinely cache *inedible* items that are not later recovered. This behavior begins within a few days after fledging, and continues through at least 8–11 months of age. Ravens older than about four months also make a hole in the ground with the bill, into which the food or other object is placed. The following example illustrates this kind of 'play caching': A young bird (8–9 months) in full view of two other young ravens, carried a stick (approximately 30 cm long), lay it down, punched a hole into the thick icy crust on the snow, then picked up the stick and inserted it end-first into the hole, pushing it in until it was out of sight. It then picked up a nearby chunk of ice and covered the hole. During food caching by adults, ravens typically avoid visual contact with conspecifics (by flying some distance away before caching) and will defend their caches if approached. In Maine, for example, adult ravens caching food typically fly off several hundred meters before hiding the food (B. Heinrich unpubl. obs.). In the example above however, two other young ravens were within a meter of the caching bird, and the object was inedible. This behavior could therefore be considered an incomplete sequence of acts performed out of context, and therefore conforms to Ficken's (1977) definition of play.

Flight play

Soon after the first feathers appear, young ravens begin wing flapping at or near the edge of the nest. At one cliff nest in Vermont, the seven youngsters (an unusually large clutch size) took turns standing at the very edge of the nest, facing inwards, flapping their wings vigorously.

Fig. 2.3. (a) play-bathing. (b) flight play, in this case wing-dipping. (c) sliding down an incline.

In one 8 minute long filmed sequence, we observed 7 bouts of flapping, each 1–2 second duration, and 4 bouts, each 8–10 second long. During these wing-beating bouts, wing-beat frequencies were higher than normally used in flight. Bouts of wing flapping increase in frequency just prior to fledging. Wing flapping while grasping the nest edge is clearly not functional in terms of locomotion, but may serve to strengthen muscles and develop motor coordination needed later in flight.

After fledging, the birds engage in frequent and vigorous 'play flights' around their natal territory, reminiscent of young horses prancing around a meadow. By their first fall, they may dive and turn by tucking in one wing, only to spread the wings out again and turn back. Sometimes they roll over in mid-flight repeatedly. They also chase each other vigorously, 'dive bombing' and feinting in and out of each other's way. Drack (1994) reported that play chasing is one of the ravens principal forms of play at a wildlife park in Austria.

Young adult ravens (probably at the age where pair formation begins), sometimes congregate in groups and engage in flight aerobatics. These 'group sky dances' occur throughout the year, but are most common in the fall among groups of juveniles (Marzluff et al. 1996). Within the broader context of the group, pairings are evident. Pairs roll and tumble and soar together. Sometimes a third bird cuts in on a displaying pair, but it is difficult to determine if pair switches occur. Mated pairs sometimes engage in graceful synchronized aerobatic displays on their own in late winter or early spring, near the onset of the breeding season.

David Lidstone (pers. comm.) in Western Maine twice saw one bird fly directly under another, fold one wing up and flip over on its back, and then lock talons with the bird on top. Both then glided with outstretched wings in this position directly into a headwind, once for about 18 seconds. David Snow (pers. comm.) observed similar behavior in ravens from Bandelier Park, in New Mexico. Demonstration of superior flying ability may be an important factor in mate choice in ravens (see review in Heinrich 1989) and we can only guess that early practice in the form of 'play' may be essential to adult success when flying ability is at a premium.

Chasing occurs in adulthood in many contexts. Adult pairs frequently chase intruders off their territories. Adults also chase each other when one tries to get a food item carried by another. In the Northwest Territories of Canada, one of us (BH) observed strings of up to a dozen birds chasing each other. In some, but not all cases, the lead bird carried an object in its bill or feet. In one case the item being carried turned out to be a loop of wire. These chases often lasted more than 10 minutes.

Bathing

When captive young ravens take their first bath in a shallow pan of water, they begin by approaching and dipping their bills, then splashing the bill back and forth in the water. They then walk into the water, hesitatingly dipping their heads, then lowering themselves into the water and beating their wings at the same time. Splashing is accompanied by 'comfort sounds'. The birds eventually hop out of the pan, seek a perch, shake violently and preen for a very short time before hopping back into the water and repeat the performance several times. In one filmed session, two fledglings alternately bathed, rolled in the soil near the pan, went through identical bathing motions in nearby grass, preened, and vocalized softly. In contrast, experienced adult birds bathe quickly and then immediately set about preening until dry. In young birds, getting clean is or could be a by-product of the activity. Ultimately, they may learn through this activity that it feels pleasant on a hot day, or cleanses the feathers.

Ravens engage in similar bathing in snow. Unlike bathing in water, however, the birds often push themselves forward through the snow on their bellies with their bills pushed into the snow, plowing ahead of them. They also sometimes stick their heads deep into the snow while flapping the wings and roll onto their backs.

Sliding down inclines

Observers from Alaskan and Northern Canadian towns routinely reported to us seeing ravens slide down steep snow covered roofs, only to fly or walk back up and repeat the slide. Ravens in our Maine aviary also roll down mounds of snow, and even do so on their backs with a stick held in the feet! David Lidstone, observing ravens at a deer carcass in Maine during the first snow storm of the year, reported that 'at least three birds flew up to a stump on a 2–3 m incline, and then slid down the slope on their backs. Twice the sliding bird was holding a stick in its talons'. Gwinner (1966) reported seeing his captive ravens repeatedly sliding down a board. We see no obvious utilitarian function for sliding behavior. Perhaps it is a social display (not necessarily play) involved in securing status or mates by 'showing off' or drawing attention to themselves.

Hanging

As reported by Gwinner (1966) and Drack (1994), ravens will also hang upside down by their feet from horizontal ropes or branches. In our birds, hanging was always on the ends of flexible twigs rather than solid ones. The birds would tip over either forwards or backwards while gripping the perch. Gwinner's birds evidently extended their wings, but in ours, the wings were usually pressed to the body and the head held horizontally as the bird turned to look around. After hanging for a period of from a few seconds up to a couple of minutes, the bird simply rotated back onto the top of the perch with the help of a few wing beats, or let go with both feet, turned in mid air and landed upright on the ground. Sometimes, while hanging upside down, the bird may let go with one foot and dangle the other using it to pass a twig back and forth between bill and foot. Drack's (1994) observation involved wild birds, so this behavior is not just an artifact of captivity.

Hanging upside down was observed in a pair of young mated birds isolated in our aviary. Between January 20 and February 10, 1995, during

Fig. 2.4. Hanging upside down, sometimes with a stick held in one foot.

about 20 hours of observation prior to the pair's first nesting, the male was observed to hang upside down on 15 occasions and the female on seven occasions. Gwinner (1966) and Drack (1994) both hypothesized that hanging by the feet is a display behavior that affects the rank of birds in a juvenile crowd, in which case it may not qualify as play.

Miscellaneous games

In the Eastern Canadian Arctic, ravens hang upside down from power lines. Richard Bargen (pers. comm.) reported the following:

> This winter (1995–6), from my window overlooking the bay at Iqaluit, I observed four ravens playing in high wind for about 15 minutes on some loose wires strung between two power poles. The wires were about a foot apart and flailing sinuously in the strong wind. First one raven would try to land on the lower wire and as it moved up and down, try to grab the upper wire in its beak. Once, one of them grabbed it with its foot instead and hung on as it whipsawed up and down in the wind. It looked like the goal was simply to hang on for as long as possible, a sort of raven rodeo. As soon as the first raven was tossed off or lost its grip (after about 20 seconds, max.), the next raven would attempt the same feat.

Another 'game' observed in ravens is 'pass-the-stick' (or other object). For example, Carlisle Spencer (pers. comm.), observing ravens in Vermont reported

> 15–20 ravens were flying in an updraft. One had a stick in its bill and although the birds rarely flapped their wings, they would dive at the bird with the stick, who was 'it'. The bird would eventually drop the stick and another would catch it before it hit the ground. This bird then became 'it'. This game continued for about 45 minutes.

Lone ravens also appear to play by dropping objects and retrieving them. Rod and Amy Adams and Elliot Swarthout (pers. comm.) observed this in the Grand Canyon, where, between 1 Sept. and 20 October 1995, ravens often seemed to play in the updrafts.

> Once we saw a raven drop a rock from its mouth and successfully catch it as it fell. Another time we saw it drop a small red object and successfully catch it. Three other times we saw a raven drop an object without retrieving it. On one occasion, a large falling feather (perhaps dropped by a raven) was pursued by six ravens, though none retrieved it. The flying ravens often transferred objects from beak to feet and back again, and from one foot to two and back again to one. One raven was flying with a foot-long [300 cm] or longer stick in its feet, giving the impression of surfing with it. It twisted its body four or five times in approximately 180 degree turns. Two other ravens followed underneath. After the raven had descended about 150 [45 m] feet into the canyon, it flew back up the rim and over a

plastic owl decoy placed in a Juniper tree to attract hawks. Once over the owl, the raven dropped the stick, which fell about 15 feet [5 metres] from the owl.

Potentially the latter behavior has the obvious utilitarian function of molesting the owl or of showing off for status or mates. However, not all objects are dropped in flight or onto potential enemies. The same observers continue:

We also saw a raven perched near the edge of the canyon pick up a small rock with one foot and then by a combination of dragging and hopping, carry it to the edge where it used it's bill to roll the rock over the edge. Then it looked down. The rock fell to a ledge 20 feet [7 metres] down. The raven then immediately repeated the whole performance.

We see no obvious utilitarian purpose, proximate or ultimate, for this solitary behavior.

Allospecific interactions

Reports, and photographs abound of ravens approaching wild predators (wolves, eagles, etc.) at a carcass, and yanking their tails. This does not result in displacing the predator from the carcass, but rather it is invariably the raven that flies up out of the way. At a game park in Austria, Drack (1994) photographed ravens pecking resting deer in the rear, and riding on top of wild boars in an area where all these species are brought into contact due to feeding by humans.

All of the hand-reared birds (some dozen) observed by BH 'harassed' carnivores (tame cats and dogs) when they first met them. Typically they approached cautiously from behind, lunged forward to peck and

Fig. 2.5. Pulling a dog's tail.

instantly jumped or flew out of the way. The young ravens became progressively bolder with each successive attempt at harrassing an animal if it did not react. Thus, an aging husky dog was eventually pecked on the nose, while a large and vigorous cat who occasionally stalked and attacked the ravens was not approached to distances less than about 4 meters. Those individual cats and dogs that were not aggressive were later completely ignored by the ravens.

The behavior of these ravens appeared playful, but ultimately may function to provide the bird with information about the intentions and responses of predators. Since ravens are unable to rip into ungulate carcasses, they rely on predators and scavengers to do so, and they are therefore brought into contact with these species. To access meat in the wild, they must therefore be able to operate in close proximity to these other species. Harassing them may be a means by which they can test and thereby gain information about the reactions of these other animals. In addition, ravens may obtain fur for lining their nests in this manner, not a trivial concern for birds that incubate at sub-zero temperatures.

Vocal play

Young ravens, even prior to fledging, engage in long and elaborate vocal monologues, unlike anything heard in adults. During this behavior, ravens produce a variety of sound types (low gurgling sounds, barely audible chortles, squeaks, quacks, loud yelling, trills and sounds that resemble water running over pebbles in a swift stream), some of which resemble calls typical of adults. These monologues are often accompanied by continually changing gestures and feather postures, as if the birds were play-acting numerous roles in an apparently random sequence. For example, loud trills and rasping quorks are accompanied by 'macho displays' – ear and throat feathers erect, and flashing nictitating membranes. In adults these displays are mating and assertive displays directed at mates or rivals. In young birds (during the first few months out of the nest), these monologues occur out of context, often either when the bird is alone or not directed at any other individual and not eliciting any obvious reaction from others. They do not appear serious since the bird may intermittently perch, stretch, yawn, pick at twigs etc., and then abruptly change to a different 'tune'.

These 'songs', like the 'subsongs' of other species of passerines, are common during the birds' first summer. In singing passerines, subsong eventually crystallizes into an adult song type. Adult ravens rarely engage in monologues, and do not have anything obviously recognizable to us as 'song'. Thus, although some of the elements of raven subsong resemble adult-like calls, there is no clear 'crystallization into adult song'.

Discussion

We have described a wide variety of raven behaviors which could, by at least some definitions, be considered 'play'. These behaviors are, in some sense, diverse and unrelated to each other except for being performed in a playful manner. However, we feel that some can be directly related to aspects of raven biology and ecology.

For example, much of the play behavior, particularly of young ravens, involves the exploration and manipulation of objects in their environment. A tendency to engage in this particular type of play makes good adaptive sense for these animals given that they live in such diverse ecological settings. Learning about what is good to eat and what is not, and what is dangerous and what is not in a 'play' context without the pressure of 'real' consequences may be critical to later success for a creature with such wide ranging distribution.

Some behaviors engaged in by ravens may be considered play: riding on boars (Drack 1994); playing 'King of the mountain' on a carcass (Heinrich 1994), or 'courtship displays' by young ravens (Lorenz 1939). On the other hand, such behaviors may also be used to signal social status. Social status plays a key role in many aspects of raven life, both among juveniles and in adults, where it appears to be central to courtship and pair formation (Gwinner 1964). Thus, it is quite possible that behaviors we might classify as 'play' in fact have a direct and important function in signalling status.

Because ravens generally mate monogamously and for life, successful competition in the mate-choice arena is absolutely critical to lifetime reproduction. Ravens engage in a number of behaviors that appear playful, but are likely to involve 'showing off' to other ravens as sexual or status enhancing displays. For example, flight aerobatics, hanging upside down, dropping and catching objects in flight etc. are all behaviors that seem playful, but may in fact have serious and immediate consequences as social displays of critical importance in establishing successful mateships.

Finally, there are clear parallels between the play of ravens and that of some mammals, whose play behaviors we are most familiar with as casual observers. For example, ravens engaging in vigorous and acrobatic play flights are reminiscent of play in young ungulates. The tendency to play 'chase' and 'keep away' with objects is reminiscent of the play of young canids, and the tendency to hang upside down and slide down inclines is reminiscent of the play of young primates. Vocal play is reminiscent of young humans. Notable however, is the lack of observations of 'play fighting' typical of many species. Young ravens engage in various forms of aggression, ranging in intensity from a quick, casual appearing feint with the bill, to grappling with talons, lunging with the bill and harsh rasping vocalizations. Although low intensity aggressive interactions sometimes appear subjectively 'playful', they may in fact be completely 'serious'. In any case, young ravens do not engage in the kind of obvious play fights that are so pervasive in young felids or canids.

Some mammals, dogs for example, have clear signals that 'label' subsequent behavior as play, and/or act as 'invitations' to play. We have not yet discerned such signals in ravens, perhaps because they do not play fight. Such signalling would seem most critical when play fighting, to ensure that all involved understand that it is not a real fight.

The process of determining what behaviors to include in this discussion of play in ravens has made us more plainly aware of the definitional problems involved in this difficult, but fascinating aspect of behavior. We hope nonetheless to have conveyed some sense of the kinds of raven behaviors that *could* be considered play to provide some basis for comparison with other species.

Play is an elusive concept. All observers of animal behavior feel that there is 'something there', but on closer examination, evades definition. In struggling with this problem whilst writing this review, we have come to think of play in terms which may be useful to other researchers. We consider behavior in general to fall along three dimensions: (1) the first corresponds to the form of motivation in terms of 'pain' vs. 'pleasure'. (2) the second dimension corresponds to how directly or indirectly the behavior is linked to ultimate fitness, spanning from very indirect with a series of proximate intermediary steps to direct with no intermediary steps. (3) the third dimension corresponds to the time delay between performance of the behavior and its ultimate functional benefit spanning from 'immediate' to 'sometime in the future'. As an example of how this can be applied, consider a young raven flapping its wings at the edge of the nest. It may be motivated to do so because it is 'pleasurable', the proximate indirect functional benefit is that it strengthens the wings for flight, and in the future, the ultimate functional benefit is that the bird enjoys higher reproductive success because of its strong flying skills, which permits it to find food more effectively or show off to mates more effectively. Perhaps all behavior can be plotted in this three dimensional space, but behavior we tend to identify as play falls on the extreme end where 'pleasure/fun' is the proximate motivation, and ultimate benefits lie in the distant future via a series of intermediary proximate benefits. As we move away from these extremes in any dimension, the behavior becomes less and less clearly defined as play. We see no clear dividing line however, between play and non-play behaviors in ravens. Rather, we see play as a concept analogous to (and perhaps correlated with) 'intelligence', in that it is a continuous quality with various dimensions, and no discrete boundaries.

References

Barrett, L., Dunbar, R. I. M., & Dunbar, P. 1992. Environmental influence on play behavior in immature Gelada baboons. *Anim. Behav.*, **44**, 111–5.

Bekoff, M. & Byers, J. A. 1981. A critical reanalysis of the ontogeny of mammalian social and locomotor play, An ethological hornet's nest. In: *Behavioral Development, The Bielefeld Interdisciplinary Project* (ed. K. Immelmann, G. W. Barlow, L. Petrinivich, & M. Main), pp. 296–337. New York: Cambridge University Press.

Bekoff, M. 1984. Social play behavior. *BioScience*, **34**, 228–33.

Caro, T. M. 1995. Short term costs and correlates of play in cheetahs. *Anim. Behav.*, **49**, 333–45.

Drack, G. 1994. Aktivitätsmuster und Spiel Freilebender Kolkraben (*Corvus corax*) im Almtal (Oberöstereich). Dissertation. Univ. Salzburg.

Fagen, R. 1981. *Animal play behavior*. New York: Oxford Univ. Press.

Ficken, M. S. 1977, Avian play. *Auk*, **94**, 573–82.

Gwinner, E. 1964. Untersuchungen uber das Ausdrucks und Sozialverhalten des Kolkraben (*Corvus corax*). *Z. Tierpsychol.*, **21**, 656–784.

Gwinner, E. 1966. Über einige Bewegungsspiele des Kolkraben (*Corvus corax* L.) *Z Tierpsychol.*, **23**, 28–36.

Heinrich, B. 1988. Why do ravens fear their food? *The Condor*, **90**, 950–2.

Heinrich, B. 1989. *Ravens in Winter*. New York: Summit Books.

Heinrich, B. 1994. Dominance and weight changes in the common raven *Corvus corax*. *Anim. Behav*, **48**, 1463–5.

Heinrich, B. 1995. Neophilia and exploration in juvenile common ravens, *Corvus corax*. *Anim. Behav*, **50**, 695–704.

Heinrich, B., Marzluff, J. & Adams, W. 1996. Fear and food recognition in naive common ravens, *Corvus corax*. *Auk*, **112**, 499–503.

Lorenz, K. 1939. Die Paarbildung beim Kolkraben. *Z. Tierpsychol*, **3**, 278–92.

Martin, P. & Caro, T. M. 1985. On the functions of play and its role in behavioral development. *Adv. Stud. Behav.*, **15**, 59–103.

Marzluff, J. M, Heinrich, B., & Marzluff, C. S. 1996. Raven roosts are mobile information centres. *Anim. Behav.*, **51**, 89–103.

Ortega, J. C. & Bekoff, M. 1987. Avian play: comparative evolutionary and development trends. *Auk*, **104**, 338–41.

Pellis, S. M. 1981a. Exploration and play in the behavioral development of the Australian magpie *Gymnarhina tihicen*. *Bird Behavior*, **3**, 37–42.

Pellis, S. M. 1981b. A description of social play by the Australian magpie *Gymnarhina tihicen* based on Eshkol-Wachmen notation. *Bird Behavior*, **3**, 61–79.

3

Object play by adult animals

SARAH L. HALL

Anthrozoology Institute, School of Biological Sciences, Bassett Crescent East, Southampton SO16 7PX, UK

What is object play?

Object play is the involvement of inanimate objects of various kinds in an animal's play activities. Fagen (1981) defines 'divertive interactions with an inanimate object ... including exploratory manipulation', when describing object play. It is a type of play behaviour familiar to pet cat and dog owners who regularly provide their pets with toys. In the absence of human manufactured toys, domestic, captive and wild animals may use a wide variety of objects such as sticks, rocks, leaves, fruit, feathers, dead prey animals and even items of discarded human bric-a-brac as objects to play with (see Heinrich & Smolker, Chapter 2). Along with other types of play described in other chapters, object play is characterised by boisterousness and appears, anthropomorphically, to be enjoyable. Objects are thrown into the air, chased and captured, pulled to pieces, kicked, shaken and bitten.

Animals of all ages, from diverse taxa, play with objects. While non-human primates and carnivorous species such as cats, dogs and bears, are regular performers, other animals also play with objects. Play with twigs, small stones and nutshells by corvids, parrots and other birds is well known (see Heinrich & Smolker, Chapter 2), as is play with human manufactured toys by a wide variety of zoo animals, from turtles to rhinoceroses.

Detailed scientific study aimed specifically at object play has been confined to a small range of species, particularly domestic cats and dogs, some wild canids, laboratory rats, primates, polecats and their domestic counter-parts, ferrets. Domestic species are convenient subjects, because they are easy to house and handle, and are less sensitive to stress induced by captivity. This is an important consideration in the study of play; animals suffering from stress are less likely to play (McCune 1992).

This is not to say that object play has not been widely studied, but that social play is usually the focus of research in playful species, since it is usually the most obvious form of play and is generally performed more frequently and for longer periods than object play.

Before introducing object play by adults, a brief discussion of object play by juveniles will serve as a worthwhile comparison.

Object play by juveniles

Studies of the function of object play by juvenile animals presume that play must have either immediate or long term effects (or both) on developing skills and behaviour. The most common hypotheses concerning the function of object play by juveniles note the structural similarity of object play to various adult behaviours, and then suggest that the juvenile, incompletely formed play version may be practice for adult performance of the same behaviour patterns in their true, 'serious' context. The main hypotheses are that juvenile object play may serve as practice for predatory behaviour (in predator species); or for tool use, particularly in non-human primates, but also in various non-primate species; and may enable exploration of stimuli, objects and environments which are new to juveniles (Ferron 1975, McGrew 1977, Sylva 1977, Beck 1980, Fagen 1981, Russell 1990, Westergaard 1992). It is also possible that juvenile object play has a part in general motor training with various physiological benefits and in the practice of interactive use of motor and perceptual skills in all species that perform it (Fagen 1976, West 1977, Smith 1982, Martin & Caro 1985, Byers & Walker 1995). Generally stated, play with objects gives juveniles opportunities to learn and to practice skills which will be vital to them when they are adults. Despite the intuitively acceptable nature of these hypotheses little direct supporting evidence has been found. For example, lack of juvenile object play in domestic kittens does not result in an adult cat with inferior, undeveloped predatory skills. In fact, kittens reared without opportunities for object play show no difference in later predatory skills when compared with cats reared with plenty of object play opportunities (Caro 1980).

Object play by adults

The focus of this chapter is object play by adult animals, a type of play behaviour that is less commonly observed and perhaps even more resistant to explanation. It is also likely to be a behaviour which is distinct from its juvenile counterpart in terms of motivation and function, despite misleading similarities in structure and external stimuli that cause it. There are fewer examples of object play by adult animals, and most are from domestic or captive animals.

Structure and function

Observation of object play by adult animals leaves most people unable to readily associate it with any particular function or even a definite structure. Generally, all types of adult play behaviour are associated with 'high arousal' behaviours; animals rarely incorporate behaviour patterns from maintenance or everyday activities, such as grazing, or grooming, into play (Barber 1991). This high arousal associated with play is easily expressed anthropomorphically. For example dwarf mongooses squeak 'excitedly' when playing (Rasa 1984). Object play by adults, despite their maturity, appears to be as boisterous and divertive for the performer as is the juvenile version.

The structure of adult object play has been described by several authors in terms of numerous rules which are applicable to all types of play (for example, Bekoff 1976, Fagen 1981). The basic components of the most commonly cited definitions include the performance of randomly ordered behaviour patterns from other types of behaviour (for example, predatory behaviour) which are performed out of context. It also involves repetition and elaboration of some, inexplicably favoured behaviour patterns directed towards an inanimate object. Such a structuralist approach to defining play predominated in earlier play literature and is much less problematic than definition by function (Loizos 1966). However, object play by adults fails to comply with some of the rules defining play; patterns from more than one type of behaviour are rarely mixed, and although there is often much repetition of particular behaviour patterns, they are often not performed randomly, but in an order similar to the 'serious' behaviour play resembles. It is often difficult to distinguish object play from the corresponding serious behaviour. For example, object play by adult cats is structurally similar to predatory behaviour.

It is much more difficult to define adult object play by function, especially because it is most commonly studied in domestic animals in which questions of function and adaptiveness are trickier to tackle. Because all types of adult play appear to lack any immediate goal or at least the function that the behaviour patterns have in the structurally similar 'serious' behaviour, adult play is easily regarded as superfluous and unnecessary (Symons 1978). It has even been denounced as 'behavioural fat', implying that it is an indulgence on the part of the animal to spend time and energy playing (Müller-Schwarze et al. 1982). In early, negative definitions, play was thought to occur only when no other behaviours, more vital to survival, were in operation, and was described, almost in passing, as a way of filling in spare time. Arousal models held that play served merely to maintain optimal levels of sensory stimulation, with no motivation being specifically for play performance (Hutt 1979).

A more sophisticated theory which is reminiscent of these ideas has been presented by Burghardt (1988), as the Surplus Resource Theory. This explains the occurrence of play with consideration of the evolution of metabolic strategies providing surplus energy resources available for play in some species, especially when juvenile. Possession of surplus resources are calculated according to various criteria including size, basal metabolic rate, level of parental care and altricial or precocial offspring. Burghardt then made nineteen predictions for the occurrence of play behaviour throughout the animal kingdom. For example, animals which are not constantly active near their physiological limits, which are not in nutritional stress and which have a large body size (reducing the per kilo cost of locomotion), are more likely to play. Where extended parental care gives juveniles time for 'boredom' play is also common. Burghardt also suggested that well-cared for captive animals could be expected to play more than wild animals since they are more likely to have the surplus energy required to spend time playing. This model sidesteps the need for direct biological benefits of play, by explaining it as a facultative response to nutritional conditions; as an exaptation (Barber 1991). This theory can be applied to all types of play, by juveniles and adults, but may be more obviously applicable to domestic and zoo animals which have ample time to spare, have no survival concerns and may need to burn up excess energy.

In contrast to negative definitions of the function of object play by adults, there have been attempts at a more constructive functional description. First, it is conventional to treat adult play as a distinct behaviour from object play by juveniles, since it is assumed that behaviour performed by juveniles has a developmental role which would be unnecessary for adults. The performance of any kind of play (object, social or locomotor) by adults presents different problems to juvenile object play when possible functions are considered.

The functional hypotheses assigned to object play by juveniles may not be so readily applied to adults. The practice hypothesis is less credible, since adult animals have generally mastered all types of behaviour necessary for their survival, making the need for practice in play unnecessary. An adult animal could practice a particular type of behaviour simply by performing it, rather than playing, unless the costs of performing the behaviour 'seriously' are higher than play (Loizos 1966, Biben 1979). However, although the costs of play have been calculated, and are small, the costs of any of the corresponding 'serious' behaviours have not (Martin 1984a, Caro 1995). It can only be assumed that predation is a more costly behaviour to perform than object play. Also, although object play by adults may include behaviour patterns used in other behaviours (for example, predatory behaviour) they can be performed in a different order, and with emphasis on favoured patterns. Thus its value as practice is questionable. However it is possible that adult object play may affect overall skill proficiency (Poole 1966, Fagen 1981). Play enables the adult to practice an already perfected basic motor skill while adding variation to the behaviour. Accordingly, play would be expected to appear in animals which have a larger and more complex behavioural repertoire, necessitating the perfection of more skills in the animals social, predatory or other behaviours (Bateson & Young 1979).

Exploration is a function often suggested for object play in adult predatory species such as domestic cats. The animal can explore and learn about novel stimuli in its environment through object play and also become familiar with objects which may initially evoke fear (Leyhausen 1979, Martin 1984b).

It is also suggested that object or manipulative play may promote tool use skills in non-human primates. Smith (1982) found that the species differences in object play in non-human primates seemed to covary with the adaptive significance of tool use for different species, and suggested that object play may have a short term facilitatory role in tool use. Both object play and tool use are more common in the Anthropoidea, and especially so in chimpanzees (McGrew 1977). In the absence of experimental evidence to support this role, others have stressed that it is possible that manipulative tool use skills may promote object play, rather than the other way round (Fedigan 1972, Chalmers & Locke-Haydon 1984). It is also possible that object play is merely a 'spin-off' from high intelligence.

The final major possible functional hypothesis is concerned with the behavioural flexibility of individuals. Fagen (1982) suggested that behavioural flexibility is an important factor in the evolution and persistence of play behaviour. He defined flexibility as the capacity of an animal to alter its behaviour in novel situations with the result of increased fitness. Animals that play might be better able to adapt skills to new requirements. Criticism of this hypothesis attacks the necessity of such versatility with respect to evolution. However, the costs of such play compared with the resulting flexibility may not be viable, and actual examples where animals have shown advantageous flexibility and any subsequent innovations in their behaviour are rare in non-human species (Smith 1982). Also, play in some animals is more stereotyped and less exaggerated than corresponding non-play behaviour (Hill & Bekoff 1977). There is no experimental evidence that directly supports the flexibility hypothesis (a common problem for play hypotheses in general).

Object play by adult predatory animals

Adult object play and predatory behaviour have structural similarities and are elicited by similar external stimuli. This suggests an association of some kind, functional or motivational, which goes beyond mere coincidence. To have a function play must have effects; for example, an individual animal's success as a predator might be enhanced by a reduced fear of novel prey and reduced time to kill, brought about by practice with prey-like objects (Russell 1990).

Most studies aimed at adult object play have used domestic carnivorous species as their subjects, necessitating an important caveat concerning use of the term function. Numerous authors have defined function, Martin & Caro (1985) defining it as 'the consequences of a behaviour which currently increase the individual's chances of survival and reproduction in the natural environment and upon which natural selection acts to maintain that behaviour'. For domestic animals the effects of behaviour on reproductive success are not relevant, because the animals are not subject to natural selection. Therefore the function of their behaviour is less significant, and has less influence on their survival. Some domestic species, such as cats, often exist in a state of semi-independence from humans, making function a slightly more relevant term (Kerby & Macdonald 1988).

A small number of studies with domestic animal subjects have provided evidence for the relation between object play and predatory behaviour (including predatory play with the prey, and the normal predation components). Sen Gupta (1988) studied object play of ferrets and domestic dogs, and concluded that there was no difference between object play and predation in structure or motivation. Perhaps object play can be distinguished from true predation by the absence of the killbite, a firm bite delivered to the object, which in nature kills the prey (Egan 1971, Bekoff 1976). Egan suggested that the lack of necessary stimuli from inanimate objects causes the loss of the killbite. However, Sen Gupta (1988) recorded the performance of the killbite during object play in both ferrets and dogs. However, whether there is a killbite or not in the play repertoire appears to be a matter of the observer's personal preference, with the result that detection of a killbite is subjective. It sometimes appears that the distinction between object play and predation is more of a human invention, especially when the distinction is narrowed down to the absence or presence of the killbite. Subjective human definition of the behaviour being performed may be a result of the presence or absence of a prey animal; an inanimate object in place of a live prey animal changes the human label from predatory behaviour to object play (Symons 1978, Smith 1982).

There are also various arguments against the importance of object play as practice for predation which have a more scientific basis. For example, Biben (1979) suggested a different means of linkage between object play and predatory behaviour in adults. She found that the probability of a kill could be predicted with known hunger and prey size. Conflict between hunger and prey size (for example, high hunger level, but large and, therefore, fear-evoking prey) caused cats to play, before, after, or instead of killing. This play she categorized as inhibited play, resulting from fear or fatigue in the cat. This play behaviour could have several functions: (1) it could enable a cat unfamiliar with the particular prey at hand to become familiar with it through the time spent in close contact with it; (2) it could increase motivation to kill the prey; (3) it could enable a fearful cat to assess the strength and possible aggression of the prey and to overcome its fear, (4) it may enable a cat to frighten and subdue a fierce and aggressive prey animal such as a rat, reducing the likelihood of the cat getting injured. Martin (1984b) also suggested that play with prey has more than one function – including, exploration, practice and inhibition of fear.

If object play, rather than predatory behaviour, is performed because objects lack a consummatory stimulus possessed by real prey (Leyhausen 1979, Martin 1984b), it is unclear why cats play with live prey. One possibility is that the animal lacks the internal motivational cues which may link hunting and killing in a hungry cat. If the cat is well-fed it may still hunt but not be motivated to kill because it is not hungry, resulting in play with live prey. This behaviour is a frustrating enigma to those interested in adult play, and one that I cannot attempt to explain here. On a similar line, West (1979) noted that play resembles behaviour with captured rather than uncaptured prey. Since hunting implies lack of capture, play cannot be relevant to this part of predatory behaviour. She continues to state that the best practice for hunting and killing is probably hunting and killing themselves, and that play with captured prey is too 'functionally ambiguous' to assign a definite theory.

Experimental evidence for the association of predation and object play in adults

A simple experimental approach to tackling the problems outlined above, based upon the observation that object play by adult domestic cats appears to be structurally related to predation, could be to test the suggestion that predation and play are causally and motivationally related (Leyhausen 1979, Hall 1995). This is necessary to eliminate the possibility of a merely coincidental similarity in structure (Caro 1995). By considering possible common causal factors which underlie performance of both types of behaviour, it may be possible to refine hypotheses about definition and function of object play. Study of play behaviour suffers from a neglect of non-evolutionary causes. Very little emphasis is placed upon proximate causes and mechanisms. Although it is unwise to apply a functional or evolutionary explanation to the presence of object play in domestic species, the tractability of domestic species presents an ideal opportunity for the study of the causal factors of play. Recent criticisms of the use of domestic animals as subjects in behaviour research have included questioning the assumption that the motivational basis of behaviour has remained unchanged throughout the reduced selection represented by domestication. However, domestic cats are a separate case. Throughout the history of the domestic cat are long periods of feral living when they were not favoured as pets (for example during the Middle Ages and Victorian era in Britain). As a result of these feral periods, in the cat's recent evolutionary history, it is unlikely that the motivational mechanisms of present day domestic cat behaviour have become dissociated from the evolutionary pressures that shaped them, especially those of predatory behaviour (Bradshaw 1992).

A number of studies have established that object play by adults (and juveniles) is elicited by objects which possess prey-like stimuli, such as size, texture, shape, odour and movement. Russell (1990) used dummies to present a range of stimuli, both prey-like and totally artificial, to ferrets, and established that stimuli which were most prey-like elicited most play. Adult domestic cats also play most intensively with toys which incorporated prey-like stimuli such as small (mouse) size, rapid movement, and a complex, furry texture (Hall 1995). Wild predator species have also been found to prefer objects with some basic prey-like characteristics as play objects. For example, Rasa (1971) found that Northern elephant seals preferred to play with small, moving objects, and suggested that this might be explained by the animals's 'search image' for small, moving prey (fish and squid).

The stimuli that elicit object play and those that elicit predation are similar. But is there a similarity in internal causal factors? In two experimental studies, I answer this question (Hall 1995, Hall & Bradshaw in prep, Hall et al. in prep). In the first, I examined the influence of hunger upon the performance of object play by cats with no hunting experience, and compared my results with those of an earlier study by Biben (1979) who studied the influence of hunger on the performance of predatory behaviour of experienced hunting cats.

Hunger is commonly used as a method of increasing an animal's motivation to perform various behaviours, including to kill prey (for example, Cabanac 1985, Ewert 1987). It also has a facilitatory effect upon predatory behaviour in other predator species. For example, sparrow hawks kill more prey, more quickly, when hungry than when sated (Mueller 1973).

In Biben's study, the level of hunger of the cats and the prey size and vulnerability were manipulated. Cats were presented with prey 0, 24 and 48 hours after their last meal. Prey were baby mice, adult mice, and young rats (cats do not commonly tackle fully grown adult rats (Childs 1986)). At the lower levels of hunger the cats were inhibited when tackling larger, stronger prey, and were less likely to kill them. However, there was a significant increase in killing with increased hunger, and an accompanying decrease in the latency to kill. Small prey were dispatched more efficiently when the cats were hungry, indicated by an increase in the incidence of the killbite (which Biben used as an indicator of intensive predatory behaviour), and in other close contact patterns such as pawing. It was possible for Biben to predict 95% of the variability in the cats' killing response if the two parameters, hunger level and prey size, were known. Biben concluded that hunger had a direct effect on the performance of predatory behaviour.

In my own study I substituted toys of two sizes, small (mouse size) and large (young rat size) for prey. These were presented to cats that either had just fed or had not eaten for 16 hours. The cats played with the small toy more intensively overall, and played more 16 hours after their last meal than immediately after eating. After 16 hours of food deprivation the small toy received more close contact behaviour patterns of a predatory nature, such as killbites. The large toy received an increase in specific, exploratory behaviour patterns such as sniffing, but the cats did not play with it. These qualitative changes in play response were specific enough to rule out the possibility that the increase in hunger simply caused an increase in the cats' general arousal, since only behaviour patterns which would be used in an encounter with prey increased in intensity with hunger. Had hunger merely caused an increase in general arousal, an increase in the intensity of all play patterns and overall locomotion would have been observed.

I suggest that the similarities in the effect of hunger on both predatory and object play behaviour indicates that the two have a shared motivational basis. Object play by adult domestic cats may not be distinguishable (at a motivational level) from predatory behaviour. Cats playing with objects may actually be hunting them. Thus, object play in non-hunting cats may be predation which has not been refined by experience to enable the cats to respond to only live prey with specific characteristics. According to this proposition it is not surprising that the expression of object play is influenced by hunger in the same way as predation in hunting cats.

My second study, again using adult domestic cats as subjects, attempted to explore the ways in which motivation for play may change, and how it could control object play. What causes an animal to stop playing with an object? There are three conventional explanations. First, motivation to play may dwindle with time. Second, other incompatible behaviours take over because their motivation has increased relative to that of play. Third, there may be consummatory behaviour patterns that end play when they are performed. It is also conceivable that a combination of these factors may stop play (Hughes & Duncan 1988).

Cats usually stop playing with individual objects after a very short time, suggesting that habituation may be a fourth, important factor in the control of play. In a series of experiments with three groups of cats, I ascertained that cats do indeed habituate to the same, unchanging object (in this case a $7 \times 5 \times 1$ cm, rectangular stuffed toy covered with white fake fur), if they are presented with the toy for three successive play sessions of a few minutes duration each. Play intensity showed a steady decline over the three play sessions. When I gave the cats a different toy (exactly the same, but black), in a fourth session, object play reappeared.

Intensity of this disinhibited play was influenced by two factors, (1) the length of the delay between each of the four play sessions, and (2) the sensory value of the toys used. If there was a short delay between play sessions, the intensity of play was greater than it was initially, in session one. Conversely, if the delay between play sessions was long, play performance was reduced in intensity compared with play in the first session.

The second, and most relevant, influence on this system of control by habituation is the overall sensory value of the toys. Using the same experimental procedure, toys of apparently higher sensory value than fake fur were used; one was covered with real roe deer fur, the other with haberdasher's 6 cm long maribou feathers. Real fur elicited more intensive and more slowly habituating play than feathers. When real fur followed three play sessions with feathers a rebound in the play response resulted, play being more intensive than it had been initially. However, when feathers followed real fur, habituation from sessions one to three was insignificant, and there was a reduction, rather than an increase, in play intensity.

These two factors, delay and toy sensory value, add variation to the central control mechanism of play by habituation summarized here. This control system is based upon two components. Firstly, the behavioural default is to habituate to a toy after initial interest and play. Once the cat has habituated it will not play unless there is a change in toy stimuli. However, this pattern is influenced by the sensory value of the toys, which modifies the effect of habituation and disinhibition on play performance.

The nature of this object play control system may be explained by comparison with predatory behaviour. In predatory behaviour any objects (including prey) which possess particular stimulus characteristics associated with prey, such as small size, complex, furry texture and rapid movement, attract the attention of the predator and can elicit play as well as predation (Rasa 1973, Leyhausen 1979, Russell 1990). Rapid habituation could prevent unnecessary performance of predatory behaviour with unsuitable prey or inanimate objects. I propose that there could be two mechanisms involved. The first is sensitivity to particular object characteristics. Possession of any of these characteristics by an object stalls habituation, and enables further play/predation to proceed. The second is sensitivity to the object's 'changeability'. In predation, the cat habituates to a prey animal that does not change with its predatory efforts. Lack of change in the prey may be because it is strong or large enough to resist the cat's predatory attempts. Despite possessing the qualities to which the first mechanism is sensitive, habituation still occurs. But if the prey is changed, that is, physically damaged, the cat will not habituate and will continue with predatory behaviour.

These two studies illustrate that the investigation of proximate causation may inform hypotheses about the role of object play in adult animal repertoires. The suggestion that object play by adult cats is homologous with predation also highlights that it may not be necessary to search for a unique function of object play by adults.

Future study of object play by adults

It may not be as important to ask what benefits an adult animal can obtain from object play, as it is for juvenile play. Rather the relevant questions may be concerned with discerning why adult animals include object play in their behavioural repertoires when it is not a behaviour with any apparent survival value. This shifts the emphasis for study from identifying a function or role, to a more basic investigation of the factors which cause and motivate an adult animal to play with an object. This change in emphasis corresponds with a recent renewal of interest in the proximate causes of play, both internal and external, an area which has received much less attention in past research of any type of play. Questions of motivation may be more accessible to experimental investigation, an important consideration for play research, which has more often relied upon correlation and optimal design studies. Experimental studies of play behaviour have proved difficult to design, but may be able to show more successfully the effects and role of play.

A further consideration may be to determine whether captive and domestic animals are the only animals that commonly perform object play as adults, and thus whether the whole behaviour is dependent upon human intervention of some kind, and is an artefact of domestication. For example, it is possible for a domestic animal, such as a cat, to survive and reproduce successfully without ever having hunted live prey, just inanimate objects and toys. Object play may have become a distinctive part of adult behaviour because of reduced opportunities for hunting. This may be a result of lack of hunting experience in an individual's lifetime and/or genetic changes in the central organization of predatory behaviour, brought about by domestication.

The most likely explanation for the performance of object play by adult predatory animals lies in its association with predatory behaviour, with which it is structurally similar and may share a motivational basis. Further studies are required to determine the extent of this association.

However, one important consideration which should be borne in mind is that object play by adults, along with all other types of play, in both adults and juveniles, does not constitute one category of behaviour. Not only do the various forms of play have different structure, causation and function, but one type of play can also differ in causation and function depending on the species performing it.

The study of object play by adult animals has practical applications. For example, knowing what factors cause and motivate object play enables it to be employed in the design of behavioural enrichment programmes for captive wild animals. Such programmes are designed to reduce the amount of time that an animal spends either being inactive or performing stereotypies. This is particularly relevant to animals kept as zoo exhibits, in which animals have very limited control over their environment, and where exhibits must be interesting for the public. Predatory species kept in zoos cannot perform their normal predatory behaviour. However, if object play is strongly associated with predatory behaviour on at least a motivational basis, then the provision of objects to play with and to 'hunt' may enable the animals to express the predatory behaviour they are motivated to perform, without offending the public. Such enrichment also exercises the animals and enables easier evaluation of their well-being (Markowitz & LaForse 1987, Mellen et al. 1981).

Acknowledgements

My research was supported by a BBSRC grant and sponsorship from WALTHAM. I would like to thank John Bradshaw for Ph.D supervision, forbearance and for constructive criticism of an earlier draft of this chapter; and Stuart Church and Mike Mendl for valuable comments on my Ph.D work.

References

Barber, N. 1991. Play and energy regulation in mammals. *Q. Rev. Biol.*, **66**, 129–47.

Bateson, P. P. G. & Young, M. 1979. The influence of male kittens on the object play of their female siblings. *Behav. Neural Biol.*, **27**, 374–8.

Beck, B. 1980. *Animal tool behavior: The use and manufacture of tools by animals*. New York: Garland STPM Press.

Bekoff, M. 1976. Animal play; problems and perspectives. In: *Perspectives in ethology Vol 2.* (eds. P. P. G. Bateson & P. H. Klopfer), pp. 165–88. New York: Plenum Press.

Biben, M. 1979. Predation and play behaviour of domestic cats. *Anim. Behav.*, **27**, 81–94.

Bradshaw, J. W. S. 1992. *The behaviour of the domestic cat*. Wallingford, Oxon: C.A.B. International.

Burghardt, G. M. 1988. Precocity, play and the ectotherm-endotherm transition. *Behaviour*, **36**, 246–57.

Byers, J. A. & Walker, C. 1995. Refining the motor training hypothesis for the evolution of play. *Amer. Nat.*, **146**, 25–40.

Cabanac, M. 1985. Influence of food and water deprivation on the behaviour of the white rat foraging in a hostile environment. *Physiol. & Behav.*, **35**, 701–9.

Caro, T. M. 1980. The effects of experience on the predatory patterns of cats. *Behav. Neural Biol.*, **29**, 1–28.

Caro, T. M. 1995. Short-term costs and correlates of play in cheetahs. *Anim. Behav.*, **49**, 333–45.

Chalmers, N. R. & Locke-Haydon, J. 1984. Correlations among measures of playfulness and skilfullness in captive common marmosets *Callithrix jacchus jacchus*. *Dev. Psychobiol.*, **17**, 191–208.

Childs, J. E. 1986. Size-dependent predation on rats (*Rattus norvegicus*) by house cats (*Felis catus*) in an urban setting. *J. Mammal.*, **67**, 196–9.

Egan, J. 1971. Object play in cats. In: *Play*. (ed. J.S. Bruner), pp. 161–5. Charmondsworth, Middlesex: Penguin Books.

Ewert, J-P. 1987. Neuroethology of releasing mechanisms: prey catching in toads. *Behav. Brain Sci.*, **10**, 337–405.

Fagen, R. 1976. Exercise, play and physical training in animals. In: *Perspectives in ethology Vol. 2.* (eds. P. P. G. Bateson & P. H. Klopfer), p. 189. New York: Plenum Press.

Fagen, R. 1981. *Animal play behaviour*. New York: Oxford University Press.

Fagen, R. 1982. Evolutionary issues in the development of behavioural flexibility. In: *Perspectives in ethology Vol 5.* (Eds. P. P. G. Bateson & P. H. Klopfer). New York: Plenum Press.

Fedigan, L. 1972. Social and solitary play in a colony of vervet monkeys *Cercopithecus aethiops*. *Primates*, **13**, 47–64.

Ferron, J. 1975. Solitary play of the red squirrel *Tamiascuirus hudsonicus*. *Can. J. Zool.*, **53**, 1495–9.

Hall, S. L. 1995. *Object play in the adult domestic cat Felis silvestris catus*. Ph.D thesis, University of Southampton.

Hill, H. L. & Bekoff, M. 1977. The variability of some motor components of social play and agonistic behaviour in infant eastern coyotes *Canis latrans*, var. *Anim. Behav.*, **25**, 907–9.

Hughes, B. O. & Duncan, I. J. H. 1988. The notion of ethological 'need', models of motivation and animal welfare. *Anim. Behav.*, **36**, 1697–707.

Hutt, C. 1979. Exploration and play. In: *Play and learning*. (ed. B. Sutton-Smith), pp. 175–94. New York: Gardner Press.

Kerby, G & Macdonald, D. W. 1988. Cat society and the consequences of colony size. In: *The domestic cat, the biology of its behaviour*. (eds. D. C. Turner & P. Bateson), pp. 67–82. Cambridge: Cambridge University Press.

Leyhausen, P. 1979. *Cat behaviour: The predatory and social behaviour of domestic and wild cats*, pp. 261–88. New York: STMP Press.

Liozos, C. 1966. Play in mammals. *Symp. Zool. Soc. Lond.*, **18**, 1–9.

McCune, S. 1992. *Temperament and the welfare of caged cats*. Ph.D thesis, University of Cambridge.

McGrew, W. C. 1977. Socialization and object manipulation of wild chimpanzees. In: *Primate bio-social development*. (eds. S. Chevalier-Skolnikoff & F. E. Poirier), pp. 261–88. New York: Garland Press

Markowitz, H. & LaForse, S. 1987. Artificial prey as behavioural enrichment devices for felines. *Appl. Anim. Behav. Sci.*, **18**, 31–43.

Martin, P. 1984a. The time and energy costs of play behaviour in the cat. *Z. Tierpsychol.*, **64**, 298–312.

Martin, P. 1984b. The (four) why and wherefores of play in cats: a review of functional, evolutionary, developmental and causal issues. In: *Play: in animals and humans*. (ed. P. K. Smith), pp. 71–94. Oxford: Basil Blackwell.

Martin, P. & Caro, T. M. 1985. On the functions of play and its role in behavioural development. *Adv. Study Behav.*, **15**, 59–103.

Mellen, J. D., Stevens, V. J. & Markowitz, H. 1981. Environmental enrichment for servals, Indian elephants and Canadian otters (*Felis serval, Elephas maximus, Lutra canadensis*) at Washington Park Zoo, Portland. *International Zoo Year Book*, **21**, 196–201.

Mueller, H. C. 1973. The devlopment of prey recognition and predatory behaviour in the American kestrel *Falco sparverius*. *Behaviour*, **49**, 313–24.

Müller-Schwarze, D., Stagge, B. & Muller-Schwarze, C. 1982. Play behaviour: Persistence, decrease and energetic compensation during food shortage in deer fawns. *Science*, **215**, 85–7.

Poole, T. B. 1966. Aggressive play in polecats. *Symp. Zool. Soc. Lond.*, **18**, 23–44.

Rasa, O. A. E. 1971. Social interaction and object manipulation in weaned pups of the northern elephant seal *Mirounga angustirostris*. *Z. Tierpsychol.*, **29**, 82–102.

Rasa, O. A. E. 1973. Prey capture, feeding techniques and their ontogeny in the African dwarf mongoose. *Z. Tierpsychol.*, **32**, 449–68.

Rasa, O. A. E. 1984. A motivational analysis of object play in juvenile dwarf mongooses *Helogale undulata rufula*. *Anim. Behav.*, **32**, 579–89.

Russell, J. 1990. *Is object play in young carnivores practise for predation?* Ph.D thesis. Senate House library, University College London.

Sen Gupta, A. 1988. *The structure and devlopment of play in ferrets and dogs*. Ph.D thesis. Senate House library, University College London.

Smith, P. K. 1982. Does play matter? functional and evolutionary aspects of animal and human play. *Behav. Brain Sci.*, **5**, 139–84.

Sylva, K. 1977. Play and learning. In: *Biology of play*. (eds. B. Tizard & D. Harvey), pp. 59–73. PA: Heinman, London 7 Lippincott.

Symons, D. 1978. *Play and aggression. A study of rhesus monkeys*. New York: Columbia University Press.

West, M. J. 1977. Exploration and play with objects in domestic kittens. *Dev. Psychobiol.*, **10**, 53–7.

West, M. J. 1979. Play in domestic kittens. In: *The analysis of social interactions*. (ed. R. B. Cairns). Hillsdale, NJ: Lawrence Erlbaum.

Westergaard, G. C. 1992. Object manipulation and the use of tools by infant baboons (*Papio cynocephalus anubis*). *J. Comp. Psychol.*, **4**, 398–403.

4

Kangaroos at play: play behaviour in the Macropodoidea

DUNCAN M. WATSON
University of Western Sydney, Hawkesbury, School of Horticulture. Richmond, New South Wales, Australia, 2753.

Introduction

The Macropodoidea (kangaroos, wallabies and rat-kangaroos) are the most distinctive and widely recognised of Australia's unique native fauna. All macropodoids are fundamentally similar in body form (Flannery 1984) but they nevertheless display an amazing diversity of species, habitat preferences and lifestyles. There are over 60 extant species [new species are still being discovered (Flannery et al. 1995) and active speciation in rock-wallabies (*Petrogale*) has led to considerable flux in their taxonomy (Eldridge & Close 1992)] divided into two families: the Potoroidae (rat-kangaroos, bettongs and potoroos) and the more derived Macropodidae ('true' wallabies and kangaroos). There are small (< 1 kg), solitary, homomorphic rainforest species (e.g. Musky Rat-kangaroo, *Hypsiprymnodon moschatus*) and large (> 90 kg), gregarious, heteromorphic species of the open arid and semi-arid plains (Red Kangaroo, *Macropus rufus*). There are arboreal tree-kangaroos (*Dendrolagus*) in the rainforests of northern Australia and New Guinea, a bettong that burrows like a rabbit (Burrowing Bettong, *Bettongia lesueur*) and species adapted to rocky outcrops and escarpments (rock-wallabies, *Petrogale* and *Peradorcas*, and Common Wallaroo, *Macropus robustus*). The range of adaptations in macropodoids was considered by Flannery (1989) to be greater than that of any placental family or superfamily with the exception of murids.

The closest placental equivalent to macropodoids are the Artiodactyla, particularly the Cervidae and Bovidae. Despite some 130 million years of phyletic separation, macropodoids and artiodactyls display striking behavioural and ecological convergence as comparisons of Jarman (1974) and Kaufmann (1974a, b) demonstrate. Macropodoids and artiodactyls represent alternative evolutionary pathways for large-bodied herbivorous mammals and this offers excellent opportunities for phylogenetic comparisons of play. Similarities and differences in the adaptive responses of macropodoids and artiodactyls and the effect on their play ought to provide independent tests of functional hypotheses of play and provide clues as to the major adaptive steps in its evolution.

Unfortunately, data on the play behaviour of macropodoids (and other marsupials) is scarce. The main aim of this review is to summarise and compile the existing data on play in the Macropodoidea. The first part of the review discusses the quality of the available data. Data quality affects the limits to which it can be confidently used to answer functional and evolutionary questions. The generally poor quality of the data for macropodoids explains why functional and evolutionary generalisations regarding play are currently premature and sets a basis for understanding why there may be confusion as to what is and what is not play in macropodoids. Following this is a review of play in macropodoids: what types of play occur and their main structural elements. The relationship between play and real fighting is examined in the third part of the review. Addressed here are the observations that some of the ritualised fights described in the literature often include characteristic features of play seen in placentals. Rather than reflecting a taxonomic difference in the fighting behaviour between macropodoids and placentals, it is argued that some of these fights were misclassified play fights. Finally, the function and evolution of play in macropodoids are discussed.

Comments on the play database of Macropodoidea

Table 4.1 is a study-by-study summary of macropodoids in which play has been reported. Where data in postgraduate theses have been published, only the published information is quoted because gaining access to these theses can be extraordinarily difficult.

Play as a category of behaviour in Macropodoidea

Play is not well-documented in macropodoids. It is generally perceived to be rare compared to placentals (e.g. Kaufmann 1974a) and an insignificant component of their behavioural repertoires. Even so, its absence has been specifically noted only twice. Vujcich (1979) did not observe any type of play in free-ranging Parma Wallabies, *Macropus parma* even though social and solitary play occurred in a sympatric population of Tammar Wallabies, *Macropus eugenii*. Volger (1976) noted that social play was absent in captive Tammar Wallabies. Only rarely is play in macropodoids defined and it is usually taken for granted that the motor patterns described were play. Kaufmann (1974a) and Croft (1981a) defined play as behaviour that did not appear to serve the adaptive purpose or, at least, not the same adaptive purpose normally assigned to similar adult behaviour. The use of the term 'adult' rather than 'functional' is rather unfortunate as it implies that adults do not play and that apparently purposeless behaviour in adults is presumably functional behaviour for which the consummatory act was not seen. Comprehensive definitions based on structural and functional criteria (see Fagen 1981) are only available for play fighting in Red-necked Wallabies, *Macropus rufogriseus* (Watson 1993; Watson & Croft 1993, 1996) and Red Kangaroos (Croft & Snaith 1991), and for locomotor play in Red-necked Wallabies (Watson 1990). Given their behavioural similarities (see Coulson 1989; Ganslosser 1989) these definitions ought to be of general use for at least other *Macropus* species.

In some studies the type of play observed might not be mentioned (e.g. 'mutual play': Hornsby 1978), or motor patterns suggestive of play may be called something other than play (e.g. 'exploratory dashes': Johnson 1987) or left unlabelled (e.g. Sanson et al. 1985). Rather than re-interpret the observations of others, Table 4.1 includes only studies in which the term 'play' was specifically used to describe particular motor patterns. Even so, mention will be given in the appropriate parts of the review of behaviour that was play-like based on contextual and structural criteria.

Study limitations

It is likely that Table 4.1 does not accurately portray taxonomic differences in play among the Macropodoidea. In reviewing the social behaviour of macropodoids, Coulson (1989) noted that the behavioural repertoires of some species (e.g. Eastern Grey Kangaroo, *Macropus giganteus* and Red Kangaroo) had been surveyed a number of times and were quite comprehensive, while little or no information was available for entire macropodid genera (e.g. *Lagorchestes*, *Lagostrophus*, *Onychogalea*, and *Dorcopsulus*) and for potoroids generally. Macropodoid research has tended to concentrate on a few species: medium to large, common species that either inhabit open habitats or are easily maintained in captivity. Information on the behavioural repertoires of the many small, rare or cryptic macropodoids, in contrast, is lacking. Such poor coverage limits the identification of evolutionary trends in macropodoid play.

Table 4.1. *Play in the Macropodoidea.*

Genus	Species	Type of play							Site[3]	Social structure[4]	Time (h)	Sample time[5]	Source
		Unspecified	Social			Nonsocial		Inter-specific					
			Fighting	Sexual	Parallel	Locomotor	Object						
POTOROIDAE													
Aepyprymnus	*rufescens*	•[1]							C	1 aM, 5 aF, 2 jF	55	N[7]	Ganslosser & Fuchs (1988)
						•			C			N[7]	Lissowsky-Fischer (1991)
Bettongia	*penicillata*				•	•			C				Lissowsky-Fischer (1991)
Potorous	*tridactylus*					•			C				Dzillum (1990)
MACROPODIDAE													
Dorcopsis	*luctuosa*		•						C	8		LE +EM	Ganslosser & Holz (1984)
			•						C	1 aM, 5 aF ; 1 aM, 3 aF; 0–1 aM, 3–6 aF; 1–4 jM; 1–2 jF	˜400	D[6]	Schappert (1984); Ganslosser & Schappert (1985)
			•						C				Ganslosser & Fuchs (1988)
Dendrolagus	*dorianus*		•						C	1 aM, 2–4 aF, 0–3 j	200	EM	Ganslosser (1983)
			•		•	•	•		C	7	˜200	D+N	Ganslosser (1979)
	ursinus		•	•					C	several aF and j; 1–2 sM		D	Schneider (1954)
	lumholtzi					•			C			D+N	Procter-Gray (1985)
Petrogale	*xanthopus*	•[1]							C	2 aF, 2j		D	Hornsby (1978)
	assimilis		•			•			F		184	EM +LE	Barker (1990)
					•				F		334	D+N	Horsup (1986)
	penicillata	•[1]							C + F			D	Hornsby (1978)
	lateralis		•			•			C			D	Engel (1985)
Thylogale	*billardierii*		•			•	•		C	2 aM, 4 aF, 1 jF	96	D+N	Clancy (1982)
			•			•			C	3 aF, 1 aM, 2 jF, 1 jM	220[5]	D	Ganslosser & Werner (1984)
						•			C				Gäng (1985)
			•						C				Köhler (1989)
	thetis	•[2]							C	4 jF, 4 jM, 8 aF		D	Wright (1991)
	stigmatica		•			•			F		247	D	Cooke (1979)
Setonix	*brachyurus*	•							C	1 aM, 1 aF			Winkler (1965)
Wallabia	*bicolor*		•			•	•		C	4aM, 5aF, 3jM	224	EM +LE	Crebbin (1982)
				•					C			D	Steck (1984)

Table 4.1. (*cont.*)

Macropus	*agilis*				•		•		C				Gäng (1985)
	eugenii		•						C	4 aF, 4 j	450	D	Russell (1973)
			•						F		120	D+N	Kinloch (1973)
			•			•			F			LE	Vujcich (1979)
			•				•		C				Russell (1989)
			•						C				Volger (1990)
	dorsalis		•						C	1 aM, 6 aF	35	M	Heathcote (1989)
	parma	•[2]							C	3 jF, 3 jM, 6 aF		D	Wright (1991)
	parryi		•	•		•	•		F		~3024	D	Kaufmann (1974a)
	rufogriseus	•							C				Winkler (1965)
			•	•					F			D	Johnson (1987)
			•	•	•	•		•	C	5–8 F; 9–11 M: variable a and j	970	EM +LE	Watson (1993); Watson & Croft (1993, 1996); Watson (pers. obs.)
					•				C				Jones (1985)
			•						C			D	Dörsum (1983)
	fuliginosus		•						C				Medd (1985)
	giganteus					•			C + F	C: 1 aM, 6 aF	C: 360 F:137	D+N	Grant (1974)
			•			•	•		F			EM +LE	Jaremovic (1984)
			•		•	•	•		F			D	Kaufmann (1975)
			•			•			F			D	Stuart-Dick (1987)
			•			•			F		~45	D	Hermann (1971)
			•			•			F				Breedon & Breedon (1966)
			•						F				Johnson (1955)
	antilopinus		•						F			D	Croft (1982)
	robustus		•			•	•		F		1600	D	Croft (1981b)
			•			•			C + F	C: 6 aM, 6 aF	846	D	Osuzawa (1978)
			•			•			F			EM +LE	Kappes (1990)
	rufus		•			•		•	F		1500	D	Croft (1981a)
			•						C + F	C: 6 aM, 5 aF, 2 jM, 2 jF		D	Croft & Snaith (1991)
						•	•		C			D	Krönert (1991)
			•						C			EM +LE	Winkler (1965)
			•	•	•	•	•	•	C	1 aM, 3 aF, 2 jF, 4 aM, 4 aF	422	D	Ganslosser & Wilhelm (1986)
			•						C	2 aF, 2 j	110	D	Russell (1973)

[1] Social play. [2] Probably play fighting and locomotor play. [3] F, field; C, captive. [4] a, adult; j, pouch young, juvenile or subadult; F, female; M, male. [5] D, day; N, night; EM, early morning; LE, late evening; M, late morning and early afternoon. [6] from Coulson (1989); [7]Udo Gorslosset (pers commun.)

Studies also differ widely in a number of methodological factors that impact repertoire completeness: the duration and timing of samples, study conditions (captive or field, and the quality of captive housing), and study group social structure (see Coulson 1989). Differences in sampling times between studies means that it may not be possible to directly compare, for example, rates of behaviour between them. The number of different types of play recorded in field and captive studies is a linear function of observation time (Fig. 4.1). Separate regressions fitted to field and captive data were not parallel ($F_{1,20} = 6.71$, $p = 0.02$) and suggest that captive studies are a more efficient means of studying play than are field studies. Even so, care must be taken in interpreting the results of captive studies as the conditions of captivity may severely limit the expression of certain types of play. Small enclosures inhibit locomotor play (Watson 1990), the absence of novel objects removes opportunities for object-related play (Wood-Gush & Vestergaard 1991) and a lack of play-partners or partners of suitable age and sex affects social play (Fagen 1981; Biben 1989). Captive studies of macropodoids do not always give adequate descriptions of housing conditions and social group characteristics so it is impossible to decide whether the absence of play or a type of play is simply a result of housing conditions or a true absence from a species' repertoire.

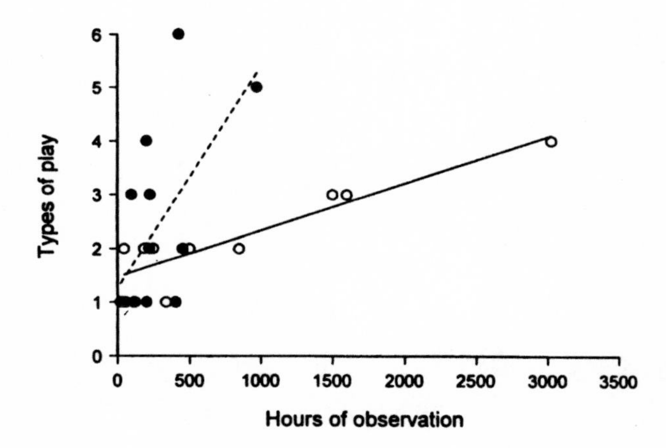

Fig. 4.1. Relationship between hours of observation and the number of different types of play observed under captive (●) and field (○) conditions. For captive conditions: $y = 1.263 + 0.004x$, $r^2 = 0.387$, $p < 0.05$; for field conditions: $y = 1.472 + 0.001x$, $r^2 = 0.800$, $p < 0.001$.

The structure of play in the Macropodoidea

Social, nonsocial and interspecific play occur in macropodoids (Table 4.1). Social play occurs as play fighting, sexual play and parallel locomotor play. Nonsocial play occurs as locomotor and object play.

Play fighting

This is the most common form of social play in macropodoids and, in terms of the range of species in which it has been recorded, is second only to locomotor play in phyletic distribution (Table 4.1). Comprehensive, quantitative analyses exist for only two species, Red-necked Wallabies (Watson 1993; Watson & Croft 1993, 1996) and Red Kangaroos (Croft & Snaith 1991). Good qualitative descriptions, including some preliminary quantification of partner preferences and rates of behaviour, are available for the Doria's Tree-kangaroo, *Dendrolagus dorianus* (Ganslosser 1979, 1983), Grey Dorcopsis Wallaby, *Dorcopsis luctuosa* (Schappert 1984), Tasmanian Pademelon, *Thylogale billardierii* (Ganslosser & Werner 1984), Whiptail Wallaby, *Macropus parryi* (Kaufmann 1974a), Eastern Grey Kangaroo (Stuart-Dick 1987), Wallaroo, *Macropus robustus robustus* (Osuzawa 1978), Euro, *Macropus robustus erubescens* (Croft 1981b) and Red Kangaroo (Croft 1981a; Wilhelm & Ganslosser 1989). Play fighting is largely mentioned anecdotally in the remaining studies.

There are a number of studies in which play fight-like behaviour were described. Sparring between females and their young were reported for the Tasmanian Pademelon (Morton & Burton 1973), Grey Dorcopsis Wallaby (Bourke 1989) and Eastern Grey Kangaroo (Jaremovic & Croft 1991). Johnson (1977) referred to 'light cuffing' of a Red-necked Pademelon, *Thylogale thetis* female's head and neck by her pouch young while Grant (1974) mentioned that 'gentle cuffing' occurred in the mother–young dyad of captive and free-ranging Eastern Grey Kangaroos. 'Mock sparring' was reported by Kinloch (1973) in juvenile Parma Wallabies. These were all probably play fights because fighting in the mother–young dyad is typically assumed to be play in macropodoids even if not explicitly stated (but see Osuzawa 1978); terms such as 'light', 'gentle' and 'mock' suggest inhibited, relaxed nonserious interactions; and, the descriptions were sometimes separate to those of real fighting suggesting that the authors considered them motivationally distinct.

Play fighting may be very common in some species. The proportion of mother–young interactions that were play fights in free-ranging animals was estimated as 33% in Euros (Croft 1981b), 11% in Red Kangaroos (Croft 1981a) and 9% in Eastern Grey Kangaroos (Jaremovic 1984). The pouch young of free-ranging Eastern Grey Kangaroos may spend 5–10% of their day play fighting (Stuart-Dick 1987). Captive animals are expected to have even higher rates of play but comparative data are unavailable. The average rate of play fighting by males in a captive group of Red-necked Wallabies was 0.124 interactions per hour at times of peak diurnal activity but rates exceeding 0.300 interactions per hour were recorded at certain times in ontogeny (Watson & Croft 1993).

Broadly, play fights involve species-typical components of fighting organised into bouts interspersed with nonfighting behaviour. The main fight motor patterns are sparring (also referred to as boxing: Croft 1981a, b; wrestling: LaFollette 1971; or grappling: Clancy 1982), pawing, cuffing, kicking and biting or mouthing. In macropodids, a distinctive high-stance posture is often adopted (see Watson & Croft 1993). During intense interactions play-partners may jump around each other in this posture, a motor pattern often referred to as skipping (Ganslosser & Wilhelm 1986; Wilhelm & Ganslosser 1989; Croft & Snaith 1991; Watson & Croft 1993). Doria's Tree-kangaroos are unusual among macropodoids in that they play fight while lying on their sides and, perhaps along with other tree-kangaroos, never kick (Ganslosser 1979). Between bouts play-partners may interact affiliatively (allogroom, touch, sniff or hold each other), feed, autogroom or simply stand vigilant. Play fights can be quite long. Some play fights by Red-necked Wallabies had over 35 bouts and lasted more than 12 minutes (Watson 1990; Watson & Croft 1996) but most lasted less than 60 seconds. The goal of play fighting may be to push an opponent towards the ground forcing it to retreat (Croft & Snaith 1991), to pin an opponent's forearms to its side and groom its throat, or to push an opponent off-balance forcing it to break contact and retreat (Watson & Croft 1993). While the initiation and termination of play fights in Red-necked Wallabies was independent of age differences between play-partners, initiators were least likely to terminate play fights (Watson 1993).

Play fights vary in intensity from slow-motion pawing to very vigorous sparring (Kaufmann 1974a; Osuzawa 1978; Batchelor 1980; Watson & Croft 1993). Watson & Croft (1993) defined three forms of play fighting in Red-necked Wallabies: refusal, and low and high intensity play fights. Refusal play fights were interactions in which an apparent invitation to play was rejected by the recipient. Low intensity play fights were those in which sparring did not occur and the main fighting pattern was pawing, while high intensity play fights were those in which sparring occurred. In addition to the defined difference between low and high intensity play fights, low intensity play fights also involved a greater rate of time spent touching, allogrooming and holding; with less time per interaction spent in bouts and in the high-stance posture; and, fewer acts per play fight and per minute.

No unambiguous play signals have been demonstrated in macropodoid play fights. Watson & Croft (1993) suggested that skipping, side and chest autogrooming and head shaking were potentially play signals in Red-necked Wallabies because of the contexts in which they occurred or because they seemed to be performed in an exaggerated and inappropriate manner. None of them were necessary for play to occur and only one, skipping, fulfilled one of the criteria of a good play signal (Bekoff 1974, 1975) by being unique to play (see also Ganslosser & Wilhelm 1986). Skipping resembled the bouncy gaits seen in the play of placentals (e.g. Bekoff 1974; Wemmer & Fleming 1974) and birds (Pellis 1981a). The combination of side and chest autogrooming with the high-stance posture also seemed to be unique to play fights. The communicative role, if any, of skipping and side and chest autogrooming was uncertain but they may be used to invite playful contact. Head shaking was more often performed by the older partner in interactions between males from different age classes (Watson & Croft 1996) and may signal the older partner's playful intentions and encourage participation by its younger partner. Head and body shaking has been suggested as play signals in placentals (e.g. Wilson & Kleiman 1974; Berger 1980). Inhibited tail-biting may serve to invite play in Doria's Tree-kangaroo (Ganslosser 1979; Ganslosser & Schappert 1985). Sniffing and allogrooming during play fights (Croft 1982; Steck 1984; Ganslosser & Wilhelm 1986; Heathcote 1989; Wilhelm & Ganslosser 1989; Watson & Croft 1993) may also have a role in play communication, but Watson & Croft (1993) could not identify any specific part of the body that was sniffed or groomed that would suggest the use of particular scent glands. As may be the case in some placentals (e.g. Black Bears, *Ursus americanus*: Henry & Herrero 1974), the prominent, mobile ears of macropodoids are well suited for serving a role in play communication (Watson & Croft 1993). Auditory signals are absent from Red-necked Wallaby play fights (Watson & Croft 1993) but Watson (pers. obs.) heard captive subadult Eastern Grey Kangaroos emit coughing vocalisations during play fights.

Self-handicapping and restraint in the performance of potentially damaging acts are features of macropodoid play fights. In interactions between mismatched partners, most obviously between females and their offspring, the advantaged partner may spend little or no time in the high-stance posture, engage in gentle defensive pawing rather than offensive sparring, and rarely if ever kick (Kaufmann 1974a, 1975; Stuart-Dick 1987; Croft & Snaith 1991; Wilhelm & Ganslosser 1989; Watson & Croft 1996). In sharp contrast, the disadvantaged partner is often the most vigorous participant. They tend to spar rather than paw, spend more time in the high-stance posture and frequently kick (Croft 1981a, b; Russell 1973; Stuart-Dick 1987; Wilhelm & Ganslosser 1989; Croft & Snaith 1991; Watson & Croft 1996). Watson & Croft (1996) found that the play fight strategy (offensive or defensive) adopted by a male Red-necked Wallaby was dependent on its partner's age. Wallabies were more likely to adopt an offensive strategy if their partner was in an older age class than themselves and adopt a defensive strategy if in a younger one. Despite these compromises, play fights between wallabies in the same age class tended to be of longer duration (i.e. more stable) and have more frequent role reversals (i.e. greater mutuality) than those of partners from different age classes. A similar pattern of self-handicapping and restraint occurs in Red Kangaroos (Croft & Snaith 1991).

Play fighting appears very early in macropodoid development, soon after young animals make their first tentative excursion from the pouch (Russell 1973; Stuart-Dick 1987; Watson & Croft 1993). Evidence from Eastern Grey Kangaroos (Stuart-Dick 1987) and Red-necked Wallabies (Watson & Croft 1993) suggests that play fight development may be tetramodal rather than bimodal as in placentals (e.g. Cheney 1978; Sachs & Harris 1978; Pfeifer 1985). Peaks in play fighting occur during the pouch young, young-at-foot, subadult and perhaps during early adulthood with troughs coinciding with the times of major developmental events: permanent vacation of pouch, weaning and sexual maturity. Watson & Croft (1993), however, failed to directly associate a decrease in play fighting at 14–16 months of age with weaning in Red-necked Wallabies. However, they argued that weaning and sexual maturity may be associated with maturational changes in play fighting and proposed three phases in its development in post-pouchlife wallabies. The first phase occurred prior to weaning and was characterized by high intensity play fights which increased in frequency and duration with age. The second phase, between weaning to sexual maturity, was distinguished by increasingly unstable play fights of short duration and a high proportion of refusal play fights. The final phase began soon after sexual maturity and was characterised by the proportion of play fights of low intensity increasing with age until, in large adult males they were the most common form of play fight seen.

A macropodoid's first play-partner is its mother and play fights outside this dyad are often noted (e.g. Kaufmann 1974a, 1975; Croft 1981a, b). Even so, play fights have commonly been reported between unrelated juveniles of both sexes (Osuzawa 1978; Ganslosser & Fuchs 1988; Schappert 1984; Watson 1993), between adult females (Heathcote 1989), between juveniles and unrelated adults of both sexes (Ganslosser 1979, 1983; Clancy 1982; Ganslosser & Schappert 1985; Ganslosser & Wilhelm 1986; Wilhelm & Ganslosser 1989; Barker 1990; Watson 1993); or virtually confined to young animals (Schappert 1984). Play fighting is generally more common in males than in females (Osuzawa 1978; Croft 1981b; Watson & Croft 1993) but not in young-at-foot Eastern Grey Kangaroos (Stuart-Dick 1987). Depending on the species, females may play more with their sons than with daughters (Croft 1981a, b; Wright 1991), equally with their sons and daughters (Wright 1991) or more with daughters than with sons (Stuart-Dick 1987). Play fights may occur more often between animals of similar dominance status (Osuzawa 1978; Watson 1993), between nonrelatives rather than half-sibs and with partners who are of slightly higher dominance status (Croft & Snaith 1991), and more often with partners of the same sex and of similar age (Watson 1993). While play between adult males and juveniles seems to be rare in most species, adult male Doria's Tree-kangaroos played with juveniles more often than did adult females, but adult females were more likely to play with nonrelatives than were adult males (Ganslosser 1983). Play fights between adult females are especially rare in Wallaroos (Osuzawa 1978), Red Kangaroos (Croft & Snaith 1991) and Red-necked Wallabies (Watson 1993). Play fights between three or more partners occur in Red-necked Wallabies (Watson 1996) and Red Kangaroos (Croft & Snaith 1991).

Sex play

This has been reported in a only a few studies (Table 4.1). It involves acts associated with male courtship (sniffing, pawing or grasping at a partner's tail or base of tail) and copulation (mounting, pelvic thrusting and penis erection). These acts may be performed as the sole play motor patterns (e.g. Schneider 1954; Kaufmann 1974a; Steck 1984; Johnson 1987), or in close juxtaposition to (Johnson 1987) or as a component of play fighting (Watson & Croft 1993). The active partner is usually a juvenile or subadult male (Kaufmann 1974a; Ganslosser & Wilhelm 1986; Johnson 1987) but juvenile, subadult and even adult females may sometimes take the active role (Ganslosser & Wilhelm 1986; D. Watson pers. obs.). The passive partner is usually the mother of a juvenile or subadult male (Kaufmann 1974a; Johnson 1987; D. Watson pers. obs.) but can be almost any other social group member (Ganslosser & Wilhelm 1986), or a male play fight partner (Watson & Croft 1993). Sex play may be first seen soon after permanent pouch vacation (Kaufmann 1974a).

Ganslosser & Fuchs (1988) considered tail grabbing and following to be elements of social play in Rufous Bettongs, *Aepyprymnus rufescens*. However, they did not mention if these acts constituted sex play, and it is possible that they were not. It is not clear if tail grabbing is a sexual act in bettongs. It is structurally different from tail grasping or pawing as the tail is firmly gripped and pulled backwards. Ganslosser & Fuchs (1988) only saw it performed by juvenile females (no juvenile males were present) toward adult females, and neither they nor Johnson (1980) described tail grasping or pawing as a component of adult sexual behaviour. However, tail grasping is a sexual act in another potoroid, the Long-nosed Potoroo, *Potorous tridactylus* (Hughes 1962). Following is often a component of courtship in macropodoids (e.g. Kaufmann 1974a; Croft 1981a, b) including Rufous Bettongs (Johnson 1980). Yet while Ganslosser & Fuchs (1988) stated that following was performed by adult males towards adult or juvenile females and by juvenile females to adult females, it was never in a sexual context.

Parallel locomotor play

This involves two or more animals performing motor patterns similar to those seen in solitary locomotor play in a coordinated, interactive manner (see below). It has been reported in several species (Table 4.1) but is rare in them. Wilhelm & Ganslosser (1989) described 'communal running bouts' between two or more captive juvenile and subadult Red Kangaroos in which they chased each other around their enclosure and engaged in brief play fights upon meeting. It was seen once among Whiptail Wallaby pouch young (Kaufmann 1974a). In this interaction, two pouch young wallabies hopped together side-by-side and then followed one another for a short distance. Three subadult male Eastern Grey Kangaroos hopped towards each other, leaped into the air, and raced around briefly together (Kaufmann 1975). Two juvenile Allied

Rock-wallabies, *Petrogale assimilis* chased each other around for 15 seconds alternating chaser/chasee roles (Horsup 1986). Play chases around bushes occurred occasionally in free-ranging juvenile Tammar Wallabies with up to four animals becoming involved (Vujcich 1979). Social 'running play' occurs but rarely in Doria's Tree-kangaroos (Ganslosser 1979). All these examples involved juvenile animals. Play chases involving adults and juveniles were reported in captive Red-necked Wallabies (Jones 1985).

Solitary locomotor play

Solitary locomotor play is the most widespread form of play in the Macropodoidea (Table 4.1). It is the most common type of play in potoroids (Table 4.1) and in some macropodids (Kaufmann 1974a; 1975; Croft 1981a; Clancy 1982; Jaremovic 1984). Included as locomotor play were reports of 'exercise play' (Jaremovic 1984) and 'playful exercise' (Osuzawa 1978). Excluded but which were probably instances of solitary locomotor play, were reports of 'exploratory dashes' in Red-necked Wallaby pouch young (Johnson 1987), 'rapid excursions' by Tammar Wallaby and Red Kangaroo pouch young (Russell 1973), and the description by Sanson et al. (1985) of a juvenile Little Rock-wallaby, *Peradorcas concinna* making continuous circuits of a rock on which its mother was basking. Also excluded were three motor patterns described by Procter-Gray & Ganslosser (1986) in captive Lumholtz's Tree-kangaroo, *Dendrolagus lumholtzi* that were suggestive of an arboreal adapted form of solitary locomotor play. Two patterns performed by a juvenile, 'quadrapedal hanging' and 'chin-up', were seen once. Another pattern, 'springing from branch to branch', was rare in calm adults but often performed by juveniles and by animals attempting to avoid capture.

With the exception of Watson (1990) who gives a detailed quantitative description of the structure of solitary locomotor play in Red-necked Wallabies, descriptions of locomotor play in macropodoids are anecdotal. In general, the locomotor play of macropodoids mimics their predator avoidance behaviour. Bouts consist of very rapid and erratic hopping in a circular or to-and-fro path in which spontaneous rapid turns and leaps into the air may occur (Barker 1990; Watson 1990). Trees, rocks and other animals may be used as obstacles for turning around or leaping over (Cooke 1979; Watson 1990). Bouts often begin and end at the same location. For juveniles this is usually their mothers (Grant 1974; Osuzawa 1978; Watson 1990) but in captive animals it may be a corner of their enclosure (Watson 1990). Bouts were interspersed with periods of relatively stationary activity such as autogrooming, feeding or vigilance. Although the locomotor play in Red-necked Wallabies was usually silent, pouch young sometimes emitted a clucking vocalisation similar to their distress vocalisation when accidently separated from their mothers (Watson 1990). This vocalisation did not have the usual consequences of a distress vocalisation: the pouch young approaching or being approached by its mother or some other adult female.

Motor patterns in locomotor play do not alway mimic predator avoidance behaviour. Captive Red Kangaroos sometimes climb onto logs during play (Wilhelm & Ganslosser 1989) and Tasmanian Pademelons dig and scatter dirt and leaves in a playful manner (Clancy 1982). Doria's Tree-kangaroos seldom engage in hopping play but do perform a slow backwards walk (Ganslosser 1979).

Locomotor play occurs soon after animals begin to leave the pouch (Kaufmann 1974a; Grant 1974; Osuzawa 1978; Croft 1981a, b; Watson 1990) but later in development in the Long-nosed Potoroo (Dzillum 1990). It is most common in the early stages of development, declining rapidly with age. Soon after permanent emergence from the pouch locomotor play is very rare in Tammar Wallabies (Russell 1989) and Eastern Grey Kangaroos (Grant 1974). Between 8–10 (pouch emergence) and 14–16 (weaning) months of age, the rate of locomotor play in Red-necked Wallabies declined 10-fold in males and 40-fold in females (Watson 1990). In contrast, the distance tranversed in play increases rapidly with age (Osuzawa 1978; Watson 1990).

Sex differences in locomotor play have been reported. In Brush-tailed Bettongs, *Bettongia penicillata* (Lissowsky-Fischer 1991) and Red Kangaroos (Krönert 1991, Lissowsky 1996) it was more common in juvenile males than in juvenile females but more common in juvenile female Eastern Grey Kangaroos than in juvenile males (Stuart-Dick 1987). In Red-necked Wallabies it was the only form of play seen in juvenile females and occurred, albeit rarely, in adult females (Watson 1990). The rate of locomotor play in juvenile males was similar to that of juvenile females but it was only seen in adulthood among young adult males. Locomotor play tended to be of longer duration in female than in male wallabies.

Locomotor play was sometimes associated with play fighting. Eastern Grey Kangaroo pouch young often ended bouts of locomotor play by engaging in play fights with their mothers (Grant 1974) as did the pouch young of Red-necked Wallabies (Watson 1990).

Object play

Object play is moderately widespread in macropodoids (Table 4.1) and may be a relatively common form of solitary play in both captive and free-ranging animals (Croft 1981a; Crebbin 1982). Because all descriptions of object play are anecdotal, it is not possible to determine if authors are confusing investigative behaviour with play. Though investigation and diversive play form a behavioural and sensory continuum, they are, nonetheless, functionally distinct activities (Fagen 1981). The extent and form of object play in macropodoids ought to be treated, at this stage, as tentative.

The dexterous forepaws and the mouth are used by macropodoids in object play. Under captive and field conditions they may manipulate or bite at sticks (Kaufmann 1975; Croft 1981b; Crebbin 1982; Jaremovic 1984), bark (Kaufmann 1974a) or grass stems (Kaufmann 1974a, 1975); grab at falling leaves (Ganslosser 1979) or mother's fur (Gäng 1985); tap or grab branches (Croft 1981b; Crebbin 1982); throw sticks against their chest (Kaufmann 1975); and wrestle with bushes (Croft 1981b; Krönert 1991). Captive animals have also been seen to manipulate pieces of food (Ganslosser 1979), paper and feathers (Wilhelm & Ganslosser 1989); pull and bite at hessian surrounding their enclosure (Clancy 1982); and manipulate and jump up and down on novel items accidentially left in their enclosures (D. Croft pers. comm.). Object play has only been seen in juvenile macropodoids. It was more common in young-at-foot than in pouch young Eastern Grey Kangaroos (Stuart-Dick 1987) and more common in juvenile male Red Kangaroos than in juvenile females (Krönert 1991). While there are insufficient data to suggest whether object play is a uniquely juvenile activity, captive adult Red Kangaroos and Tammar Wallabies are very unresponsive to novel stimuli and will usually ignore novel objects left in their enclosures (Russell & Pearce 1971).

Interspecific play

Unambiguous examples of interspecific play in macropodoids are rare (Table 4.1). As for object play it is possible that some accounts may be confusing investigative behaviour with play. The most obvious example of interspecific play occurs between hand-reared macropodoids and their human caregivers. It is very easy to elicit play fighting in young hand-reared Euros, Red kangaroos and Eastern Grey Kangaroos by placing a hand on their chest and gently pushing or rubbing (D. Watson pers. obs.). In wildlife parks, hand-reared or habituated subadult and young adult males (usually *Macropus*) may approach humans and initiate a play fight. In the past, this response was exploited and large male Eastern Grey and Red Kangaroos were trained to fight with humans as circus attractions (Johnson 1955).

Unambiguous examples of mutual interspecific play between nonhumans and macropodoids have not been reported. However, a captive juvenile Red-necked Wallaby was observed to repeatedly approach and follow an Australian Magpie, *Gymnorhina tibicen* for about 30 seconds (D. Watson pers. obs.). The encounter ended with both animals feeding within a few metres of each other. Both appeared relaxed and the wallaby gained no obvious resource by supplanting the bird. Australian Magpies are particularly playful birds (Pellis 1981a, b, 1983) and it is therefore possible that the interaction was mutual play. Similar interactions have been reported between Domestic Cattle, *Bos taurus* and Whiptail Wallabies (Kaufmann 1974a) and between Domestic Cattle and Eastern Grey Kangaroos (Kaufmann 1975). In these encounters, the cattle (usually a calf) took the approacher role. Kaufmann considered these to be active displacements, though he gave no indication as to how far the macropodoids were moved or if the cattle gained access to any resource. It is possible that this could have been play from at least the cattle's point of view. A potential play encounter, at least on behalf of the kangaroo, between a Red Kangaroo and an Emu *Dromaius novahollandiae* in captivity was observed by Udo Ganslosser (pers. comm.).

Nonmutual interspecific play has been reported only once. Wilhelm & Ganslosser (1989) suggested that attempts by captive juvenile Red Kangaroos to catch insects was a form of solitary play. Whether this was exploratory behaviour, play, failed predatory attempts or an artefact of captivity is uncertain. Wild adult Tammar Wallabies have been seen to actively chase, catch and eat hawkmoths (Borthwick et al. 1979).

Real fighting or play fighting?

Real fighting in the Macropodoidea has two forms: unritualised and ritualised. Unritualised fights are intense, brief encounters in which participants show very little restraint in the performance of potentially damaging acts. Mainly restricted to highly competitive situations among adult males in large macropodid species (especially access to oestrous females: Murböck 1975; Croft 1981a, b), they are the only form of fighting seen in potoroids, tree-kangaroos and between females of small to medium-sized macropodids (Ganslosser 1989). Ritualised fights, on the other hand, are stereotyped encounters in which potentially damaging tactics are avoided, at least initially. They can be of quite long duration and are the most common type of fighting seen in large male macropodoids. They are usually assumed, in the absence of empirical evidence, to have a role in the establishment and maintenance of dominance relationships (e.g. Kaufmann 1974a; Ganslosser 1989) and are thus considered to be equivalent to the ritualised fights of artiodactyls (e.g. Geist 1974; Clutton-Brock et al. 1979, 1982). A comprehensive review of the structure of fighting in macropodoids is given by Ganslosser (1989).

The structure of ritualised fights and play fights is very similar (Germann-Meyer & Schenkel 1970; Kaufmann 1974a; Grant 1974; Croft 1981a, b, 1982; Jaremovic 1984; Johnson 1987; Ganslosser & Wilhelm 1986). So what criteria have been used to tell them apart? Generally, it has often boiled down to simply the identity of participants: fights between a female and her offspring or between juveniles are assumed to be play while fights between other animals are assumed to be ritualised fights. In the latter case, the performance of fight motor patterns was assumed to imply an aggressive motivation regardless of context. This is best demonstrated in Kaufmann's (1974a) landmark study of Whiptail Wallabies, the results of which have been reiterated in almost every ethological study of macropodoids since. Play was defined as behaviour that '...apparently serve no immediate adaptive function, or at least not the same function that they serve when performed by adults' (p. 337). But ritualised fights between subadult and adult males '... conferred no immediate benefit on the winners ...' (p. 321) and occurred in the '... absence of any immediate object to be won ...' (p. 315). So, in the end, purposeless behaviour by juvenile Whiptail Wallabies was play but purposeless behaviour by subadults and adults was not! There has often been a failure among students of macropodoid behaviour to recognise that species-typical motor patterns of fighting can occur in different motivational contexts, not just aggression (e.g. Fry 1987). The performance of fight motor patterns should not be considered divorced from the context in which they occurred.

The close structural similarities between ritualised fighting and play fighting, and the lack of adequate definitions suggests the possibility that some of the ritualised fights described in the literature were misclassed play fights. This issue has been previously examined by Watson (1993), Watson & Croft (1993, 1996), and Croft & Snaith (1991). What follows is largely a compilation of their arguments. If true then play fighting in macropodoids may be more common than has previously been supposed and would explain why many descriptions of ritualised fights include structural and contextual features more suggestive of play than of aggression when contrasted with the ritualised fights and play fights of artiodactyls and other placentals. Table 4.2 lists these play-like features and the studies in which they were observed.

Macropodoids self-handicap against a smaller or weaker opponent in ritualised fights by standing flat footed and by not using their obviously superior strength when pawing, cuffing, pushing, holding and grasping. The behaviour of mismatched animals thus often appear to sharply contrast each other. Adult macropodoids may engage in gentle pawing while their juvenile partners spar vigorously and do most of the kicking (Kaufmann 1974a; Osuzawa 1978). Even among partners of similar size and strength, ritualised fights may appear relaxed interactions. Kaufmann (1974a) used terms like 'gentlemanly' and 'relaxed attitude' to describe some ritualised fights of Whiptail Wallabies. More, ritualised fights are often limited to long periods of slow-motion or low intensity pawing (LaFollette 1971; Kaufmann 1974a, 1975; Batchelor 1980) and affiliative acts like allogrooming, touching and holding often occur. Regardless of how well participants are matched in a ritualised fight, they are often interactions in which there is considerable repetition of motor patterns and a high level of mutual involvement. Both partners may actively prolong the interaction by initiating bouts, and initiation may be a role alternated between participants. Acts identified in other studies as potential play-markers (skipping, side and chest autogrooming) have been reported in ritualised fights.

The context in which ritualised fights occur is often more like that of play than agonism. They are more common in juvenile and young adult males than in large adult males; and often appear purposeless as access to some limiting resource (food, water, shelter or mates) is not identified as the reason for the interaction. Antagonists may also remain in close proximity after a fight (sometimes even feeding peacefully together) and dominance relations between participants based on the outcome of ritualised fights may be obscure [e.g. high rates of dominance reversals (Kaufmann 1974a; Osuzawa 1978; Johnson 1985) and indecisive contests (Grant 1974; Croft 1981a, b) occur] or not the same as that resulting from the outcome of clearly competitive situations or supplant/avoidance interactions.

Table 4.2. *Play-like features of macropodoid ritualised fights.*

Feature	Species	Source
STRUCTURAL		
Self-handicapping and physical restraint	*Thylogale billardierii*	Clancy (1982)
	Macropus parryi	Kaufmann (1974a)
	Macropus robustus	Osuzawa (1978)
	Macropus rufus	Croft (1981a); Croft & Snaith (1991)
Allogrooming or other affiliative acts	*Wallabia bicolor*	Crebbin (1982)
	Macropus irma	Heathcote (1989)
	Macropus parryi	Kaufmann (1974a)
	Macropus rufogriseus	LaFollette (1971)
	Macropus antilopinus	Croft (1982)
	Macropus robustus	Croft (1981b)
Aggressive acts performed in a relaxed manner	*Petrogale penicillata*	Batchelor (1980)
	Thylogale billardierii	Clancy (1982)
	Macropus eugenii	Murphy (1976)
	Macropus rufogriseus	LaFollette (1971)
	Macropus parryi	Kaufmann (1974a)
	Macropus robustus	Osuzawa (1978)
	Macropus rufus	Croft (1981a)
Repetition and mutuality	*W. bicolor*	Crebbin (1982)
	Macropus parryi	Kaufmann (1974a)
	Macropus giganteus	Kaufmann (1975; Jaremovic (1984)
	Macropus robustus	Croft (1981b)
	Macropus rufus	Croft (1981a)
Subordinate is most likely to kick	*Thylogale billardierii*	Clancy (1982)
	Macropus rufogriseus	Johnson (1989)
	Macropus parryi	Kaufmann (1974a)
	Macropus giganteus	Kaufmann (1975)
	Macropus robustus	Osuzawa (1978); Croft (1981b)
Acts associated with play (skipping, and autogrooming)	*Thylogale stigmatica*	Cooke (1979)
	Macropus rufogriseus	Fleming et al. (1983); Johnson (1985)
	Macropus parryi	Kaufmann (1974a)
	Macropus giganteus	Grant (1974); Kaufmann (1975); Southwell (1976)
	Macropus robustus	Croft (1981b)
	Macropus rufus	Croft (1981a); Ganslosser & Wilhelm (1986)
CONTEXTUAL		
Access to a resource not clearly associated with the interaction	*Macropus parryi*	Kaufmann (1974a)
	Macropus robustus	Croft (1981)
	Macropus rufus	Croft (1981a); Croft & Snaith (1991)

Table 4.2. (*cont.*)

Feature	Species	Source
Participants remain in close proximity after the interaction	*Macropus parryi*	Kaufmann (1974a)
	Macropus rufogriseus	Johnson (1989)
	Macropus giganteus	Grant (1974)
Unclear dominance relationships between participants	*Macropus parryi*	Kaufmann (1974a)
	Macropus rufogriseus	Johnson (1985); Murböck (1979)
	Macropus robustus	Osuzawa (1978)
	Macropus rufus	Croft (1981a)
Predominantly an activity of young males	*Wallabia bicolor*	Crebbin (1982)
	Macropus parryi	Kaufmann (1974a)
	Macropus giganteus	Jaremovic (1984)
	Macropus robustus	Croft (1981b)
	Macropus rufus	Croft (1981a)

If all ritualised fights in macropodoids are assumed to be equivalent to the ritualised fights of artiodactyls then there are major structural and contextual differences between them. For instance, ritualised fights in Red Deer, *Cervus elaphus* stags (Clutton-Brock et al. 1979, 1982) and the dominance fights of Mountain Sheep, *Ovis canadensis* rams (Geist 1974) are components of male rutting behaviour and so have a clear purpose: (1) give the winner unambiguous dominance over the loser; (2) are most common in adult males in their prime reproductive years; (3) do not involve affiliative acts or self-handicapping; and, (4) are not relaxed encounters. Though rams may stay in close proximity after a fight, the loser henceforth acts in an utterly subordinate manner to the winner. These differences require an explanation. Are they due to a real taxonomic difference in the behaviour of macropodoids and artiodactyls or are they due to some macropodoid play fights being misclassed as ritualised fights?

The first explanation implies that the ritualised fights of macropodoids resemble, if only superficially, play fighting in artiodactyls and other placentals because either the selective forces that have driven the evolution of their agonistic behaviour differed from that of placentals or they have responded in an alternative, but equally adaptive, way to similar forces. Differences in fighting behaviour between macropodoids and artidactyls is thus an expected, if not an inevitable, result of their long evolutionary and geographic isolation. While this is not an unreasonable hypothesis, differences in macropodoid and artiodactyl fighting have generally been taken for granted by macropodoid researchers and the evolutionary and functional implications of these differences ignored.

The main problem with this hypothesis is that while macropodoids and artiodactyls differ greatly in morphology, metabolism and reproductive biology, they also exhibit considerable convergence in response to similar selective forces. Both have evolved compartmentalised stomachs and foregut fermentation enabling them to deal efficiently with an increasingly fibrous diet (see Freudenberger et al. 1989) and there are similar adaptive interrelationships between social structure, habitat, diet, and body size and form (see Kaufmann 1974a, b; Jarman 1974; Jarman & Coulson 1989; Norbury et al. 1989; Croft 1989). More, convergence in the fundamental characteristics of ritualised fighting (and play) is evident between birds and placentals even though the evolutionary gap between them is wider than between macropodoids and artiodactyls. To assume that differences in the ritualised fighting behaviour of macropodoids and artiodactyls are due to evolutionary divergence is tantamount to arguing, without supporting data, for an exception to the general pattern of convergence in other areas of behaviour and ecology.

The alternative hypothesis, that descriptions of ritualised fights in macropodoids often appear play-like because what is being observed is play and not aggression, provides a simpler but more confronting explanation. Mistakes in classification may have a number of possible causes. Few studies of macropodoid social behaviour have looked at social relationships on an individual animal basis. Most group individuals into age and sex classes thus losing a significant amount of information on social relationships. The significance of repeated fights between the same individuals, for example, may be lost. Mistakes may also occur because of unfamiliarity with the behavioural repertoire of a species and with standard ethological practices. This is likely to occur where behavioural observations are peripheral to the main aims of a study or are but a small part of a study with much broader aims, where the observer is not ethologically trained, and where the behavioural repertoire of a species is poorly documented. Under these conditions, it is to be expected that criteria previously used to define play and nonplay fighting may be applied dogmatically.

A few studies flagged the possibility that some ritualised fights might be play (Croft 1981a, 1982; Heathcote 1989). But the issue was not critically addressed until recently (Croft & Snaith 1991; Watson 1993; Watson & Croft 1993, 1996). The working definition of play fighting in Red-necked Wallabies used by Watson & Croft (1993) and Watson (1993) and implicit in Watson & Croft (1996) was fighting in which genus-specific agonistic signals were absent and immediate access to a resource was not involved. Similarly, Croft & Snaith (1991) defined play fighting in Red Kangaroos as fights in which immediate access to a resource was not identified. (These definitions are similar to working definitions of play used in placental studies: e.g. Bekoff 1972; Chalmers 1978, 1980; Symons 1978.) Thus defined, play fighting in Red-necked Wallabies and Red Kangaroos shared many of the structural and contextual criteria of play in placentals and birds. They mixed motor patterns from different motivational contexts, exaggerated the performance of certain motor patterns, self-handicapped, exhibited physical restraint in the performance of potentially damaging acts, frequently reversed attacker/defender roles or took the same role, repeated acts and sequences of acts often and displayed high levels of reciprocity or mutuality. Compared to agonistic behaviour, the social relationships formed and ontogeny of play were markedly different. The most parsimonious conclusion, was that the behaviour observed was play and not agonism and that many, but not necessarily all, of the ritualised fights described in the literature were also play. This implies that play in the Macropodoidea might be more common than has previously been supposed and the social structure and apparent aggressiveness of macropodoids may, in some cases, need re-evaluating.

Function and evolution of play in the Macropodoidea

The Macropodoidea and Artiodactyla share very similar evolutionary histories. Both were derived from small-bodied, forest-dwellers that had selective diets and probably dispersive social organisations that, in response to similar habitat, diet and predation pressures beginning in the mid to late Miocene explosively radiated and gave rise to grazing species that live in large groups, have high levels of social cohesion, polygynous mating systems and marked sexual dimorphism in body size and form (Eisenberg 1987; Flannery 1984; 1989). While the details of adaptive change may differ due to differences in the intensity of selective pressures and to different evolutionary contingencies (*sensu* Gould 1989), the broad sweep of change remains remarkably convergent. Males in polygynous macropodoids are larger than females and have highly increased upper body musculature and forearm length to improve their fighting abilities and to provide visual cues as to their fighting capabilities. Similarly, in polygynous artiodactyls, males are larger than females but have evolved horns and antlers as visual displays and weaponry.

Tracing descent through modification, Byers (1984) argued that the ancestral form of play in the Perrisodactyla and Artiodactyla was solitary locomotor play (mainly of motor patterns used in predator avoidance) and its ancestral function (and, therefore, the ancestral function of all play) was motor training (Bekoff & Byers 1981). Social play as play fighting, he argued, evolved later because its selection required the appearance of social groups and the potential for polygyny. He proposed that the ancestral function of social play was also motor training, but that other functions may have subsequently evolved as a result of morphological and behavioural specialisation leading to modifications in play.

There are insufficient data to confidently speculate as to the function and evolution of play in macropodoids but it is reasonable to suppose a similar evolutionary trend as outlined by Byers (1984). Broadly, as patterns of behaviour and social organisation are more derived and complex in macropodoids than in potoroids so too for play. The Potoroidae are homomorphic and have not evolved polygyny so play fighting should be either rare or nonexistent. The data seem to confirm this as play fighting has not been reported in potoroids. Most of their play is locomotor or parallel locomotor activity and contact social play has only been reported in the Rufous Bettong. The apparent paucity of social play in potoroids compared to macropodoids may thus be a real taxonomic difference and not a coverage artefact, but it will not be known for certain until more complete behavioural repertoires of potoroids become available. Taxonomic trends within families, however, are not discernible. All genera of Macropodidae appear to engage in all forms of social and non-social intraspecific play when differences in genera diversity, species coverage and study conditions are controlled. Differences appear to reflect species-specific differences in lifestyle.

The degree to which predation has impacted the evolution of the Macropodoidea is a contentious issue (Jarman & Coulson 1989). Even so, macropodoids have been prey to a suite of mammalian, avian and reptilian predators throughout their evolution (see reviews by Jarman & Coulson 1989; Robertshaw & Harden 1989). For this reason, the training of predatory avoidance skills in locomotor play should be the most ubiquitous form of play in macropodoids. Table 4.1 shows this to be the case. Species-specific variations in the frequency or ontogenetic patterns of locomotor play may reflect differences in predator avoidance strategies and habitat structure. In general, potoroids rely on crypsis to avoid predators (Heinsohn 1968) as do some macropodids (e.g. Bridled Nailtailed Wallaby, *Onychogalea fraenata*; Gordon 1983). Predation pressures may also be reduced in tree or rock-dwelling species because their habitat greatly reduces the velocity of a cursorial predator's attack (Croft 1987). Reduced selective pressure for the practice of predator avoidance skills suggests, other things being equal, that locomotor play should be rarer than in species more reliant on crypsis as a predator avoidance strategy and in those species inhabiting trees or rocky habitats.

Parallel locomotor play also mimics predator avoidance behaviour and may have evolved to provide extra realism to motor training. That parallel locomotor play occurs in Brush-tailed Bettongs but not in other potoroids may be because the former are exposed to a greater predation risk than the other potoroids listed in Table 4.1. The Brush-tailed Bettong's habitat incorporates more open vegetation associations than that of either the Long-nosed Potoroo and the Rufous Bettong (see Seebeck et al. 1989) and the survival of young to adulthood is very low (Christensen 1980).

If locomotor play has been modified to serve functions other than motor training for predator avoidance, then other functions may most likely be found in macropodoids that have become highly specialised for living in trees and rocky habitats. In these animals, competence at climbing and moving safely and efficiently may have as much immediate survival value as predator avoidance. Falls, for example, may have deadly consequences (e.g. Hornsby 1973). Tree-kangaroos have a number of motor patterns adapted for movement in trees (Procter-Gray & Ganslosser 1986) and a unique scrambling motor pattern for climbing up rocks occurs in Allied Rock-wallabies (Barker 1980). The practice of climbing skills and cognitive benefits such as the learning of terrain variability (Müller-Schwarze 1984; Gomendio 1988) may be important derived functions of locomotor play. The 'chin-up' and 'quadrapedal hanging' motor patterns performed by juvenile Lumholtz's Tree-kangaroos therefore appear to be examples of locomotor play in which the function is not the practice of predatory avoidance but of motor training aimed at the development of climbing skills.

The function of play fighting in macropodoids was examined by Watson & Croft (1993, 1996), Watson (1993) and Croft & Snaith (1991). Based on design features, play-partner preferences and age/sex differences, Watson & Croft (1993, 1996) and Watson (1993) concluded that the most likely function of play fighting in Red-necked Wallabies was motor training for intraspecific fighting. However, it was clear that motor training did not provide a complete explanation of play fighting. Unless the energy expended was greater than it subjectively appeared, motor training did not seem to be the primary benefit gained from low intensity play fights, and the benefits gained by the advantaged partner in play fights between mismatched play-partners was obscure. Developmental differences in play fighting indicated that the primary benefits obtained by individuals were age-related. Whereas the play fighting of young male wallabies was best and most simply explained as motor training, that of large adult males was not. Socialisation benefits (e.g. the development and maintainance of social cohesion: Wilson 1973; Bekoff 1978; Bekoff & Byers 1981) seemed even less appropriate as other, less risky alternatives were available for achieving the same benefits (e.g. allogrooming) and large adult males are the least socially bound of all Red-necked Wallaby age and sex classes (Johnson 1989). Croft & Snaith (1991) proposed that an important function of play fighting in Red Kangaroos was to enable individuals to inform each other of their fighting capabilities and dominance status in a nonthreatening situation. Low intensity play fights may be a means by which this is achieved in Red-necked Wallabies. However, the difficulty with an assessment function is that animals are vulnerable to the communication of misinformation (Fagen 1981).

If the function of play fighting is fight motor training, then taxonomic differences in play fighting behaviour should parallel taxonomic differences in fighting behaviour. In general, fighting is more complex and derived in macropodids than in potoroids and in heteromorphic, polygynous macropodids than in homomorphic, nonpolygynous species (Ganslosser 1989). That potoroids have not yet been seen to play fight suggests their fighting is structurally so simple it requires little practice in the form of play. Unfortunately, there are insufficient data to correlate structural differences in play fighting with variations in sexual dimorphism in macropodids. Unique, species-specific fighting patterns have evolved in macropodoids. Specialisations in hindleg and tail morphology make kicking in at least a bipedal stance impossible in tree-kangaroos and they are unique among macropodids in lying on their sides to wrestle in the primitive fighting pattern of potoroids (Ganslosser 1989). The absence of kicking from fighting in Doria's Tree-kangaroo is parallelled by its absence from their play fighting, and they also play fight while lying on their sides (Ganslosser 1979).

Sex play has so far been reported in relatively social, heteromorphic and polygynous macropodids. It is likely, therefore, to be a derived form of play. There are a number of potential adaptive benefits that it may provide depending on the context in which it occurs. When performed as the sole form of play by young male macropodids it may provide motor training practice for sexual behaviour. Sex acts performed by themselves in play, however, seem to be rare in early ontogeny and stop being performed well before males have had an opportunity to engage in real sex. Unless they can maintain early training benefits through alternative behaviour in later ontogeny (e.g. performance of sex acts during play fighting), these skills may be minimal and lost or reduced by the time they can apply their skills in reproduction. A role for sex play in the development of social bonds between a mother and her son seems unlikely as, at least in the *Macropus* species in which sex play has been reported, males disperse from their natal home ranges at sexual maturity (Kaufmann 1974a; Johnson 1986; Oliver 1986) and so are unlikely to form long term relationships with their mothers. Sex acts performed in the context of a play fight may help to promote social contact (in other words, serve a similar purpose to allogrooming in play), be a means by which one animal asserts its dominance over its partner, or simply be a rather purposeless manifestation of excitement. When females adopt the male role in sex play they may do so for socialisation or dominance assertion purposes.

Tail-grabbing and following by Rufous Bettongs are unique play motor patterns in macropodoids. They are not clear components of sexual behaviour but it is possible they are modified sexual acts that have secondarily evolved a new function in play. They do not seem to closely mimic any nonplay behaviour so a motor training function seems inappropriate. A role in promoting or facilitating social bonds between females and between males and females is possible but not compelling. Both male and female bettongs are solitary but less so than other potoroids (Seebeck et al. 1989). While captive females are unaggressive towards each other and they may develop a loose social system in the wild based on familiarity with neighbours (Johnson 1980), there is no evidence of long-lasting social bonds between subadults and adult females (Ganslosser & Fuchs 1988).

Young macropodoids probably gain a number of adaptive benefits from object play. By repeatedly manipulating objects with their forepaws and mouths, pouch young develop their head and paw coordination in a manner analogous to the predatory play of avian (e.g. Pellis 1983; Watson 1992) and mammalian (e.g. Barrett & Bateson 1978; Biben 1979) predators. Another, quite motivationally distinct, training benefit is the development of skill at performing complex motor patterns used during intraspecific aggression. Playfully throwing sticks against their chests and wrestling with bushes resembles the ritualised bush displays of adult male macropodids during agonistic encounters (e.g. Kaufmann 1974a; Croft 1981a, b). A cognitive benefit may be the familiarisation with environmental features, particularly the edibility of different potential food items.

The predominantly bipedal gait of macropodoids has enabled their forelimbs, like those of primates, to become modified for tasks other than locomotion. Apart from their evolution as weapons in males of polygynous species, they are also highly specialised for selecting and manipulating food items. It is plausible, therefore, that learning to effectively use their forelimbs in feeding was the original selective pressure for the evolution of object play in macropodoids. The practice of motor patterns used in agonistic displays is most likely a derived benefit contingent upon the evolution of polygyny. Cognitive benefits such as determining the suitability of items as food is unlikely to be a function because food preferences may be obtained more directly by a pouch young observing its mother while she feeds and by sampling or sniffing the food she eats (Croft 1981a, b).

Conclusion

The Macropodoidea, and marsupials generally, are often perceived of as primitive mammals in whom sophisticated physiological, morphological, behavioural and ecological adaptations comparable to that of placentals are absent or at least of a much lower order. The only reason macropodoids evolved in Australia was assumed to be because of Australia's long geographic isolation and the absence of more advanced and ecologically competitive placental herbivores (Johnson 1955). Accumulated knowledge in palaeontology, physiology, ecology, ethology and anatomy has emphatically demolished this perception. The reality is that macropodoids display an array of adaptations equal to, and sometimes exceeding, those of any placental (Dawson 1977). The long held and cherished notion that placentals replaced marsupials everywhere except Australia because they were competitively superior is fast becoming untenable. Fossil evidence suggests that marsupials outcompeted placentals in South America and, less certainly, in Australia (Archer 1993). Feral populations of macropodoids, moreover, have and are successfully competing with placentals in many parts of the world (see reviews by Gilmore 1977 and Maynes 1989).

An outcome of this negative perception of marsupials has been the assumption that complex motor patterns like play which are considered to be characteristic of 'intelligent' animals are unlikely to occur or, if they do, to be simple in form and diversity. Russell (1974) suspected that when the social behaviour of marsupials was studied, an emphasis was placed on finding primitive characters to support preconceived notions of marsupial behaviour. That the Macropodoidea have often been thought of as unplayful may have as much to do with this preconceived bias as with unfamiliarity with their behaviour.

This review aimed primarily to debunk another myth surrounding marsupials by showing that macropodoids are indeed playful animals and engage in complex and diverse play. Their playfulness may have been grossly underestimated and their aggressiveness equally overestimated because some fights described in the literature were misclassed as serious rather than play. However, it is also a review in which the lack of qualitative and quantitative behavioural data in macropodoids, particularly of potoroids and the rarer, more cryptic macropodids was emphasised. Until more complete studies are made of the behaviour of these animals, confident predictions regarding the evolution and function of play in macropodoids and comparative analyses comparing the structure and function of play in macropodoids with their placental counterparts, the artiodactyls, cannot be made.

References

Archer, M. 1993. The murgon monster. *Aust. Nat. Hist.*, **24**, 60–1.

Barker, S. C. 1990. Behaviour and social organisation of the allied rock-wallaby *Petrogale assimilis*, Ramsey, 1877 (Marsupialia: Macropodoidea). *Aust. Wildl. Res.* **17**, 301–11.

Barrett, P. & Bateson, P. 1978. The development of play in cats. *Behaviour*, **66**, 106–20.

Batchelor, T. A. 1980. The social organisation of the brush-tailed rock wallaby (*Petrogale penicillata*) on Motatapu Island. M.Sc. Thesis. Univ. of Auckland.

Bekoff, M. 1972. The development of social interaction, play, and metacommunication in mammals: An ethological perspective. *Quart. Rev. Biol.*, **47**, 412–34.

Bekoff, M. 1974. Social play and play-soliciting by infant canids. *Amer. Zool.*, **14**, 323–40.

Bekoff, M. 1975. The communication of play intention: Are play signals functional? *Semiotica*, **15**, 231–9.

Bekoff, M. 1978. Social play: structure, function and the evolution of a co-operative social behavior. In: *The Development of Behavior*. (eds. Burghardt, G. & Bekoff, M.) New York: Garland, pp. 367–83.

Bekoff, M. & Byers, J. A. 1981. A critical reanalysis of the ontogeny and phylogeny of mammalian social play: An ethological hornet's nest. In: *Behavioural Development in Animals and Man.* (eds. Immelmann, K., Barlow, G., Main, M. & Petrinovich, L.) New York: Cambridge Univ. Press, pp. 296–337.

Berger, J. 1980. The ecology, structure and function of social play in bighorn sheep (*Ovis canadensis*). *J. Zool.* (Lond.), **192**, 531–42.

Biben, M. 1979. Predation and predatory play behavior of cats. *Anim. Behav.*, **27**, 81–94.

Biben, M. 1989. Effects of social environment on play in squirrel monkeys: resolving Harlequin's Dilemma. *Ethology*, **81**, 72–82.

Borthwick, J. A., Langworthy, R. A., & Turner, J. 1979. Omnivorous Tammars? *West. Aust. Res.*, **14**, 133.

Bourke, D. W. 1989. Observations on the behaviour of the Grey Dorcopsis Wallaby, *Dorcopsis luctuosa* (Marsupialia: Macropodidae) in captivity. In: *Kangaroos, Wallabies and Rat-kangaroos.* (eds. Grigg, G. C., Jarman, P. J. & Hume, I. D.) Chipping Norton: Surrey Beatty and Sons, pp. 633–40.

Breedon, S. & Breedon, K. 1966. *The Life of the Kangaroo.* Sydney: Angus and Robertson.

Byers, J. A. 1984. Play in ungulates. In: *Social Play in Animals and Humans.* (ed. Smith, P. K.,) Oxford: Blackwell Scientific Publ., pp. 43–65.

Chalmers, N. R. 1978. A comparison of play and non-play activities in feral olive baboons. In: *Recent Advances in Primatology Vol I: Behavior*. (eds. Chivers, D. J. & Herbert, J.) New York: Academic. Press, pp. 131–4.

Chalmers, N. R. 1980. The ontogeny of play in feral olive baboons (*Papio anubis*). *Anim. Behav.*, **28**, 570–85.

Cheney, D. L. 1978. The play partners of immature baboons. *Anim. Behav.*, **26**, 1038–50.

Christensen, P. E. S. 1980. The biology of *Bettongia penicillata* Gray, 1837, and *Macropus eugenii* Desmarest, 1804, in relation to fire. *Forests Dept. Of West. Aust., Bull.* No. 91.

Clancy, T. F. 1982. Aspects of the behaviour of *Thylogale billardierii* in captivity. B. Sc.(Hon.) Thesis, University of Tasmania, Hobart.

Clutton-Brock, T. H., Albon, S. D., Gibson, R. M. & Guinness, F. E. 1979. The logical stag: Adaptive aspects of fighting in red deer (*Cervus elaphus* L.). *Anim. Behav.*, **27**, 211–25.

Clutton-Brock, T. H., Guiness, F. E. & Albon, S. D. 1982. *Red Deer: Behaviour and Ecology of Two Sexes.*Chicago: Univ. of Chicago Press.

Cooke, B. N. 1979. Field observations of the behaviour of the macropod marsupial, *Thylogale stigmatica* (Gould). M. Sc. Thesis, Univ. of Queensland, Brisbane.

Coulson, G. 1989. Repertoires of social behaviour in the Macropodoidea. In: *Kangaroos, Wallabies and Rat-kangaroos.* (eds. Grigg, G. C., Jarman, P. J. & Hume, I. D.) Chipping Norton: Surrey Beatty and Sons, pp. 457–73.

Crebbin, A. 1982. Social organisation and behaviour of the Swamp Wallaby, *Wallabia bicolor* (Desmarest) (Marsupialia – Macropodidae) in captivity. B.Sc (Hons.) Thesis, Univ. of New South Wales, Sydney.

Croft, D. B. 1981a. Behaviour of red kangaroos (*Macropus rufus*) in northwestern New South Wales, *Australia. Aust. Mamm.*, **4**, 5–58.

Croft, D. B. 1981b. The social behaviour of the euro (*Macropus robustus*) in the Australian arid zone. *Aust. Wildl. Res.*, **8**, 13–49.

Croft, D. B. 1982. Some observations on the behaviour of the antilopine wallaroo, *Macropus antilopinus* (Marsupialia: Macropodidae). *Aust. Mamm.*, **5**, 5–13.

Croft, D. B. 1987. Socio-ecology of the antilopine wallaroo, *Macropus antilopinus*, in the Northern Territory, with observations on sympatric *M. robustus woodwardii* and *M. agilis*. *Aust. Wildl. Res.*, **14**, 243–56.

Croft, D. B. 1989. Social organization of the Macropodoidea. In: *Kangaroos, Wallabies and Rat-kangaroos*. (eds. Grigg, G. C., Jarman, P. J. & Hume, I. D.) Chipping Norton: Surrey Beatty and Sons, pp. 505–25.

Croft, D. B. & Snaith, F. 1991: Boxing in red kangaroos, *Macropus rufus*: Aggression or play? *Intern. J. Comp. Psychol.*, **4**, 221–36.

Dawson, T. J. 1977. Kangaroos. *Sci. Amer.*, **237**, 78–89.

Dorsum, K. 1983. Verhaltenbeobachtungen am Bennettkänguruh, *Macropus rufogriseus fruticus* (Ogilby 1838). Zulassungsarbeit, Universität Heidelberg.

Dzillum, E. 1990. Dominanzbeziehungen in einer gruppe von Langschnauzen Kainchenkänguruh *Potorous tridactylus* (Kerr 1792). Diplomarbeit, Universität Bielefeld.

Eisenberg. J. F. 1987. The evolutionary history of the Cervidae with special reference to the South American radiation. In: *Biology and Management of the Cervidae*. (ed. Wemmer, C. M.) London: Smithsonian Inst. Press, pp. 60–64.

Eldridge, M. D. B. & Close, R. L. 1992. Taxonomy of rock wallabies, *Petrogale* (Marsupialia: Macropodidae). I. A revision of the eastern Petrogale with the description of three new species. *Aust. J. Zool.*, **40**, 605–26.

Engel, E. 1985. Untersuchungen zum sozialverhalten des Filskänguruhs (*Petrogale lateralis*) in Simpson Gap (Australien). Diplomarbeit, Universität Heidelberg.

Fagen, R. 1981. *Animal Play Behavior*. Oxford: Oxford Univ. Press.

Flannery, T. F. 1984. Kangaroos: 15 million years of Australian bounders. In: *Vertebrate Zoogeography and Evolution in Australasia* (*Animals in Space and Time*) (eds. Archer, M. & Clayton, G.) Carlisle: Hesperian Press, pp. 817–35.

Flannery, T. F. 1989. Phylogeny of the Macropodoidea; a study of convergence. In: *Kangaroos, Wallabies and Rat-kangaroos*. (eds. Grigg, G. C., Jarman, P. J. & Hume, I. D.) Chipping Norton: Surrey Beatty and Sons, pp. 1–46.

Flannery, T. F., Boeadi, Szalay, A. L. 1995. A new tree-kangaroo (Dendrolagus: Marsupialia) from Irian Jaya, Indonesia, with notes on ethnography and the evolution of the tree-kangaroos. *Mammalia*, **59**, 65–84.

Fleming, D., Cinderey, R. N. & Hearn, J. P. 1983. The reproductive biology of Bennett's wallaby (*Macropus rufogriseus rufogriseus*) ranging free at Whipsnade Park. *J. Zool. (Lond.)*, **201**, 283–91.

Freudenberger, D. O., Wallis, I. R. & Hume, I. D. 1989. Digestive adaptations of kangaroos, wallabies and rat-kangaroos. In: *Kangaroos, Wallabies and Rat-kangaroos*. (eds. Grigg, G. C., Jarman, P. J. & Hume, I. D.) Chipping Norton: Surrey Beatty and Sons, pp. 179–87.

Fry, D. P. 1987. Differences between play fighting and serious fighting among Zapotec children. *Eth. and Sociobiol.*, **8**, 285–306.

Gang. L. 1985. Verhaltenbeobachtungen am Flinken Känguruh (*Macropus agilis* (Gould 1842)) unter besonder berücksichtigung des sozialverhaltens. Diplomarbeit, Universität Heidelberg.
Ganslosser, U. 1979. Sociale Kommunikation, Gruppenlieben, speil- und Jugendverhalten des Doria-Baumkänguruhs (*Dendrolagus dorianus* Ramsey 1833). *Z. Säugetierkd.*, **44**, 137–53.
Ganslosser, U. 1980. An annotated bibliography of social behaviour in kangaroos (Macropodidae). *Säugetierkd. Mitt.*, **28**, 138–48.
Ganslosser, U. 1983. Quantitative studies on the social behaviour of Doria's tree kangaroo (*Dendrolagus dorianus*) in captivity. I. Structure and development of a family group. *Zool. Anz.*, **211**, 1–29.
Ganslosser, U. 1989. Agonistic behaviour in Macropodoids – a review. In: *Kangaroos, Wallabies and Rat-kangaroos.* (eds. Grigg, G. C., Jarman, P. J. & Hume, I. D.) Chipping Norton: Surrey Beatty and Sons, pp. 475–503.
Ganslosser, U. & Fuchs, C. 1988. Some remarks on social behaviour in captive rufous rat-kangaroos. *Zool. Anz.*, **220**, 300–12.
Ganslosser, U. & Holz, C. 1984. Zum kampfverhalten verschiedener sozialer kategorien beim streifenbuschkänguruh (*Dorcopsis muelleri*). *Verh. Dt. Zool. Ges.*, **77**, 281.
Ganslosser, U. & Schappert, I. 1985. Soziobiologie sum socialspiel bei Doria-Baumkänguruh (*Dendrolagus dorianus*) und Striefenbuschkänguruh (*Dorcopsis muelleri*) (Macropodidae). *Verhand. Dt. Zool. Ges.*, **78**, 206.
Ganslosser, U. & Werner, I. 1984. Zur Häufigkeit freundlicher und agonistischer Verhaltensmuster in einer Gruppe von Rotbauchfilandern (*Thylogale billardieri*, Desmarest) (Marsupialia, Macropodidae). *Zool. Anz.*, **213**, 51–67.
Ganslosser, U. & Wilhelm, P. 1986. Sex- and age-dependent differences in frequencies of social interactions between captive red kangaroos, *Macropus rufus* (Desmarest, 1822). *Zool. Anz.*, **216**, 58–71.
Geist, V. 1974. *Mountain Sheep*. Chicago: Univ. of Chicago Press.
Germann-Meyer, V. & Schenkel, R. 1970. Über das Kampfverhalten des grauen Riesenkänguruhs, *Macropus giganteus. Rev. Suisse Zool.*, **77**, 938–42.
Gilmore, D. 1977. The success of marsupial as introduced species. In: *The Biology of Marsupials.* (eds. Stonehouse, B. & Gilmore, D.) London: Macmillan Press, pp. 279–348.
Gomendio, M. 1988. The development of different types of play in gazelles: implications for the nature and functions of play. *Anim. Behav.*, **36**, 825–36.
Gordon, G. 1983. Bridled Nailtail Wallaby *Onychogalea fraenata.* In: The *Australian Museum Complete Book of Australian Mammals.* (ed. Strahan, R.) Sydney: Angus & Robertson Publ., p. 205.
Gould, S. J. 1989. *Wonderful Life. The Burgess Shale and the Nature of History.* London: Penguin Books.
Grant, T. R. 1974. Observations of enclosed and free-ranging grey kangaroos (*Macropus giganteus*). *Z. f. Säugetierkd.*, **39**, 65–78.
Heathcote, C. F. 1989. Social behaviour of the Black-striped Wallaby, *Macropus dorsalis*, in captivity. In: *Kangaroos, Wallabies and Rat-kangaroos.* (eds, Grigg, G. C., Jarman, P. J. & Hume, I. D.) Chipping Norton: Surrey Beatty and Sons, pp. 625–8.

Heinsohn, G. 1968. Habitat requirements and reproductive potential of the macropod marsupial *Potorous tridactylus* in Tasmania. *Mammalia*, **32**, 30–43.

Henry, J. D. & Herrero, S. M. 1974. Social play in the American black bear: its similarities to canid social play and an examination of its identifying characteristics. *Amer. Zool.*, **14**, 371–89.

Herrmann, D. 1971. Beiobachtungen des Gruppenlebens ostaustralicher Graugrokänguruhs, *Macropus giganteus* (Zimmermann, 1777) und Bennett-känguruhs, *Protemnodon rufogrisea* (Desmarest, 1817). *Säugetierkd. Mitt.*, **19**, 352–62.

Hornsby, P. E. 1973. Some preliminary findings from the study of the behaviour of South Australian rock wallabies: the yellow-footed rock wallaby, *Petrogale xanthopus*, and the Pearson's Island rock wallaby, *Petrogale penicillata pearsonii*. 45th ANZAAS Conf. Perth, Australia. Sect. II: *Zoology*, 1–8.

Hornsby, P. E. 1978. A note on the pouch life of rock wallabies. *Vict. Nat.*, **95**, 108–11.

Horsup, A. B. 1986. The behaviour of the allied rock-wallaby, *Petrogale assimilis* (Macropodinae). B.Sc.(Hons.) Thesis, James Cook Univ. of North Queensland, Townsville.

Hughes, R. 1962. Reproduction in the macropod marsupial *Potorous tridactylus* (Kerr). *Aust. J. Zool.*, **10**, 193–224.

Jaremovic, R. 1984. Space and time related behaviour in eastern grey kangaroos (*Macropus giganteus* Shaw). Ph.D. Thesis. Univ. of New South Wales, Sydney.

Jaremovic, R. V. & Croft, D. B. 1991. Social organization of Eastern Grey Kangaroos in Southwestern New South Wales. II. Associations with mixed groups. *Mammalia*, **55**, 543–54.

Jarman, P. J. 1974. The social organisation of antelope in relation to their ecology. *Behav.*, **58**, 215–67.

Jarman, P. J. 1989. Sexual dimorphism in Macropodoidea. In: *Kangaroos, Wallabies and Rat-kangaroos*. (eds. Grigg, G. C., Jarman, P. J. & Hume, I. D.) Chipping Norton: Surrey Beatty and Sons, pp. 433–47.

Jarman, P. J. & Coulson, G. 1989. Dynamics and adaptiveness of grouping in macropods. In: *Kangaroos, Wallabies and Rat-kangaroos*. (eds. Grigg, G. C., Jarman, P. J. & Hume, I. D.) Chipping Norton: Surrey Beatty and Sons, pp. 527–47.

Johnson, C. N. 1985. Ecology, social behaviour and reproductive success in a population of red-necked wallabies. Ph.D. Thesis, Univ. of New England, Armidale.

Johnson, C. N. 1986. Philopatry, reproductive success of females and maternal investment in the red-necked wallaby. Behav. *Ecol. Sociobiol.*, **19**, 143–50.

Johnson, C. N. 1987. Relationships between mother and infant red-necked wallabies (*Macropus rufogriseus banksianus*). *Ethology*, **74**, 1–20.

Johnson, C. N. 1989. Social interactions and reproductive tactics in red-necked wallabies (*Macropus rufogriseus banksianus*). *J. Zool. (Lond.)*, **217**, 267–80.

Johnson, D. H. 1955. The Incredible Kangaroo, *Nat. Geogr.*, **108**, 487–500.

Johnson, K. A. 1977. Ecology and management of the red-necked pademelon, *Thylogale thetis*, on the Dorrigo Plateau of northern New South Wales. Ph.D. Thesis, Univ. of New England, Armidale.

Johnson, P. M. 1980. Observations on the behaviour of the rufous rat-kangaroo, *Aepyprymnus rufescens*, in captivity. *Aust. Wildl. Res.*, **7**, 347–58.

Jones, B. 1985. Zum sozialverhalten zweier zoo-populationen des Bennett-Wallaby. Diplomarbeit, Universität Erlangen.

Kappes, G. 1990. Aspekte aus dem sozialverhalten weiblicher Wallaroos *Macropus robustus robustus* (Gould 1841). Diplomarbeit, Universität Erlangen.

Kaufmann, J. H. 1974a. Social ethology of the whiptail wallaby, *Macropus parryi*, in north-eastern New South Wales. *Anim. Behav.*, **22**, 281–369.

Kaufmann, J. H. 1974b. The ecology and evolution of social organization in the kangaroo family. *Amer. Zool.*, **14**, 51–62.

Kaufmann, J. H. 1975. Field observations of the social behaviour of the eastern grey kangaroo, *Macropus giganteus*. *Anim. Behav.*, **23**, 214–21.

Kinloch, D. I. 1973. Ecology of the parma wallaby *Macropus parma* Waterhouse, 1846, and other wallabies on Kawau Island, New Zealand. M.Sc. Thesis, Univ. of Auckland.

Köhler, A. 1989. Beobachtungen zur entwicklung eines jungetieres biem Rotbauchfilander (*Thylogale billardieri*). Unpubl. Report, Institut fur Zoologie I.

Krönert, P. 1991. Öko-ethologische untersuchungen an weiblichen Roten Riesenkänguruh (*Macropus rufus*) und ihren folgejungen in Zentralaustralien. Diplomarbeit, Universität Erlangen.

LaFollette, R. M. 1971. Agonistic behaviour and dominance in confined wallabies (*Wallabia rufogriseus frutica*). *Anim. Behav.*, **19**, 93–101.

Lissowsky, M. 1996. The occurrence of play behaviour in marsupials. In: *Comparison of Marsupial and Placental Behaviour*. (eds. Croft, D. B. & Ganslosser, U.) Filander Verlag Gmbh, pp. 187–207.

Lissowsky-Fischer, M. D. 1991. Untersuchungen zur jungtierentwicklung bei *Bettongia penicillata* (Gray 1837) in gefangenschaft. Diplomarbeit, Universität Erlangen.

Maynes, G. M. 1989. Zoogeography of the Macropodoidea. In: *Kangaroos, Wallabies and Rat-kangaroos*. (eds. Grigg, G. C., Jarman, P. J. & Hume, I. D.) Chipping Norton: Surrey Beatty and Sons, pp. 47–66.

Medd, C. 1985. Social interaction and use of space by the Western Grey Kangaroo (*Macropus fuliginosus*) in captivity at Paignton Zoo. Research Project Report.

Morton, S. R. & Burton, T. C. 1973. Observations on the behaviour of the macropodid marsupial *Thylogale billardieri* in captivity. *Aust. Zool.*, **18**, 1–14.

Müller-Schwarze, D. 1984. Analysis of play behavior: What do we measure and when. In: *Social Play in Animals and Humans*. (ed. Smith, P. K.) Oxford: Blackwell Sci. Publ., pp. 147–58.

Murböck, A. 1979. Welchen Einfluß haben Gehegegröße und Tierdichte auf das Verhalten des Bennett wallabis, *Macropus rufogriseus fruticus*, Ogilby, 1838? *Säugetierkd. Mitt.*, **27**, 216–40.

Murböck, T. 1975. Zum kampfverhalten des Bennettwallabis, *Protemnodon rufogriseus frutica*, Ogilby 1838. *Säugetierkd. Mitt.*, **23**, 79–80.

Murphy, L. 1976. The social behaviour of the tammar wallaby, *Macropus eugenii* (Desmarest), in captivity (Marsupialia: Macropodidae). B.Sc.(Hons.) Thesis, Univ. of New South Wales, Sydney.

Norbury, G. L., Sanson, G. D. & Lee, A. K. 1989. Feeding ecology of the Macropodoidea. In: *Kangaroos, Wallabies and Rat-kangaroos*. (eds. Grigg, G. C., Jarman, P. J. & Hume, I. D.) Chipping Norton: Surrey Beatty and Sons, pp. 169–78.

Oliver, A. 1986. Social organisation and dispersal in the red kangaroo. Ph.D. Thesis, Murdoch Univ., Melbourne.

Osuzawa, J. E. 1978. Activity and behaviour of the wallaroo, *Macropus robustus* Gould. M. Nat. Res. Thesis, Univ. of New England, Armidale.

Pellis, S. M. 1981a. A description of social play by the Australian magpie *Gymnorhina tibicen* base on Eshkol-Wachman notation. *Bird Behav.*, **3**, 61–79.

Pellis, S. M. 1981b. Exploration and play in the behavioural development of the Australian magpie, *Gymnorhina tibicen*. *Bird Behav.*, **3**, 37–49.

Pellis, S. M. 1983. Development of head and foot co-ordination in the Australian magpie, *Gymnorhina tibicen*, and the function of play. *Bird Behav.*, **4**, 57–62.

Pfeifer, S. 1985. Sex differences in social play of scimitar-horned oryx calves (*Oryx dammah*). *Z. Tierpsychol.*, **69**, 281–92.

Procter-Gray, E. 1985. The behaviour and ecology of Lumholtz's Tree-kangaroo, *Dendrolagus lumholtzi* (Marsupialia: Macropodidae). Ph.D. Thesis, Harvard Univ., New York.

Procter-Gray, E. & Ganslosser, U. 1986. The individual behaviours of Lumholtz's tree kangaroo: repertoire and taxonomic implications. *J. Mammal.*, **67**, 343–52.

Robertshaw, J. D. & Harden, R. H. 1989. Predation on Macropodoidea: A review. In: *Kangaroos, Wallabies and Rat-kangaroos*. (eds. Grigg, G. C., Jarman, P. J. & Hume, I. D.) Chipping Norton: Surrey Beatty and Sons, pp. 735–53.

Russell, E. M. 1973. Mother-young relations and early behavioural development in marsupials, *Macropus eugenii* and *Megaleia rufa*. *Z. Tierpsychol.*, **33**, 163–203.

Russell, E. M. 1974. Recent ecological studies on Australian marsupials. *Aust. Mammal.*, **1**, 189–212.

Russell, E. M. 1989. Maternal behaviour in the Macropodoidea. In: *Kangaroos, Wallabies and Rat-kangaroos*. (eds. Grigg, G. C., Jarman, P. J. & Hume, I. D.) Chipping Norton: Surrey Beatty and Sons, pp. 549–69.

Russell, E. M. & Pearce, G. A. 1971. Exploration of novel objects by marsupials. *Behav.*, **40**, 312–22.

Sachs, B. D. & Harris, V. S. 1978. Sex differences and developmental changes in selected juvenile activities (play) of domestic lambs. *Anim. Behav.*, **26**, 678–84.

Sanson, G. D., Nelson, J. E. & Fell, P. 1985. Ecology of *Peradorcas concinna* in Arnhemland in a wet and a dry season. *Proc. Ecol. Soc. Aust.*, **13**, 65–72.

Schappert, I. 1984. Verhaltenbeobachtungen beim Buschkänguruh, *Dorcopsis muelleri*, in Zoologischen Gärten. *Vivarium Darmstadt*, **1**, 2–12.

Schneider, K. M. 1954. Vom baumkängaruh (*Dendrolagus leucogenys* Matschie). *D. Zool. Garten*, **21**, 63–106.

Seebeck, J. H., Bennett, A. F. & Scott, D. J. 1989. Ecology of the Potoroidae. In: *Kangaroos, Wallabies and Rat-kangaroos*. (eds. Grigg, G. C., Jarman, P. J. & Hume, I. D.) Chipping Norton: Surrey Beatty and Sons, pp. 67–88.

Southwell, C. 1976. Behaviour and social organisation of three species of macropod at Diamond Flat, New South Wales. B.Sc.(Hons.) Thesis, Univ. of New England, Armidale.

Steck, A. 1984. Zum sozialverhalten des Sumpf-Wallabies, *Wallabia bicolor* (Desmarest 1804). Diplomarbeit, Universität Heidelberg.

Stuart-Dick, R. I. 1987. Parental investment and rearing schedules in eastern grey kangaroos. Ph.D. Thesis, Univ. of New England, Armidale.

Symons, D. 1978. *Play and Aggression. A Study of Rhesus Monkeys.* New York: Columbia Univ. Press.

Volger, K. 1976. Beobachtungen zur entwicklung der jungtiere von Tammar-Wallabies, *Macropus eugenii* (Desmarest, 1817). Diplomarbeit, Universität Bielefeld.

Volger, K. 1990. Untersuchung zur ramlichen und sozialen struktur und deren entwicklung bei Tammer-Wallabies, *Macropus eugenii* Desmarest 1817) Dissertotion, Universität Bielfeld.

Vujcich, M. V. 1979. Aspects of the biology of the parma (*Macropus parma* Waterhouse) and darma (*Macropus eugenii* Desmarest) wallabies with particular emphasis on social organisation. M.Sc. Thesis, Univ. of Auckland, Auckland.

Watson, D. M. 1990. Play behaviour in a captive group of Red-necked Wallabies (*Macropus rufogriseus banksianus*). Ph.D. Thesis, Univ. of New South Wales, Sydney.

Watson, D. M. 1992. Object play in a laughing kookaburra *Dacelo novaeguineae*. *Emu*, **92**, 106–8.

Watson, D. M. 1993. The play associations of red-necked wallabies (*Macropus rufogriseus banksianus*) and relation to other social contexts. *Ethology*, **94**, 1–20.

Watson, D. M. & Croft, D. B. 1993. Play fighting in captive red-necked wallabies, *Macropus rufogriseus banksianus*. *Behaviour*, **126**, 219–45.

Watson, D. M. & Croft, D. B. 1996. Age-related differences in play fighting strategies of captive male red-necked wallabies (*Macropus rufogriseus banksianus*). *Ethology*, **102**, 336–46.

Wemmer, C. & Fleming, M. J. 1974. Ontogeny of playful contact in a social mongoose, the meerkat, *Suricata suricata*. *Amer. Zool.*, **14**, 427–36.

Wilhelm, P. & Ganslosser, U. 1989. Sequential organisation of social behaviour in captive adult and juvenile *Macropus rufus* (Marsupialia: Macropodidae). *Aust. Mammal.*, **12**, 5–14.

Wilson, S. 1973. The development of social behaviour in the vole (*Microtus agrestis*). *Zool. J. Linn. Soc.*, **52**, 45–62.

Wilson, S. & Kleiman, D. 1974. Eliciting play: a comparative study. *Amer. Zool.*, **14**, 341–70.

Winkler, S. 1965. Beobachtungen über den gebrauch der Heinde an verschiedenen beuteltierarten. Wiss. Hauserbeit, Universität Frankfurt.

Wood-Gush, D. G. M. & Vestergaard, K. 1991. The seeking of novelty and its relation to play. *Anim. Behav.*, **42**, 599–606.

Wright, S. 1991. Maternal investment in the Red-necked Pademelon and Parma Wallaby. B.Sc.(Hons.) Thesis, Univ. of New South Wales, Sydney.

5

Intentional communication and social play: how and why animals negotiate and agree to play

MARC BEKOFF

Department of Environmental, Population, and Organismic Biology, University of Colorado, Boulder, CO 80309–0334 USA

and COLIN ALLEN

Department of Philosophy, Texas A&M University, College Station, TX 77843 USA

Social play: evolution, pretense, and the cognitive turn

To return to our immediate subject: the lower animals, like man, manifestly feel pleasure and pain, happiness and misery. Happiness is never better exhibited than by young animals, such as puppies, kittens, lambs, etc., when playing together, like our own children. Even insects play together, as has been described by that excellent observer, P. Huber, who saw ants chasing and pretending to bite each other, like so many puppies. (*Charles Darwin 1871/1936, p. 448*)

Pierre Huber (1810, p. 148), in his book about the behavior of ants, claims that if one were not accustomed to treating insects as machines one would have trouble explaining the social behavior of ants and bees without attributing emotions to them. Although we shall skirt the issue of emotion, many observers would agree that animals play because it is fun for them to do so. But even if the issue of emotions is set aside, readers conditioned by the scruples of modern psychology are likely to be skeptical of Darwin's ready acceptance that Huber observed ants playing. Play, as the quotation above indicates, seems to involve pretense, and pretense is commonly thought to require more sophisticated intentions than are usually attributed to ants. How could Huber have seen or inferred pretense from the behavior of the ants? And how could he be sure that the observed behavior was not, in fact, directed toward some very specific and immediate function? These questions raise the difficult issue of what play is, or, as biologists are wont to put it, how to define 'play'. This issue has proven a great challenge to those who study this interesting behavioral phenotype.

We and others believe that social play is a tractable, evolved behavioral phenotype that lends itself to detailed empirical study. And, the flexibility and versatility of social play makes it a good candidate for comparative and evolutionary cognitive studies including those that center on ways in which animals might negotiate agreements to engage in a cooperative social interaction. As such, cognitive ethological approaches are useful for gaining an understanding of the social play behavior of diverse animals for a number of reasons (Jamieson & Bekoff 1993; Bekoff 1995a,b; Allen & Bekoff 1997; Bekoff 1998). First, empirical research on social play has benefited and will further benefit from a cognitive approach because play involves issues of communication, intention, role playing and cooperation. Second, many believe that detailed analysis of social play may provide more promising evidence of animal minds than research in many other areas, for it may yield clues about the ability of animals to understand each other's intentions. Third, play is a phenomenon that occurs in a wide range of mammalian species and a number of avian species. Thus it affords the opportunity for a comparative investigation of cognitive abilities extending beyond the all-too-common narrow focus on primates that dominates discussions of nonhuman cognition. Thus, the topic of social play exemplifies many of the theoretical issues faced by cognitive ethologists and may help those who are interested in broadening the evolutionary study of animal cognition.

The study of social play provides an opportunity to pursue the suggestion by Niko Tinbergen (1972) and others (Schaller & Lowther 1969) that we may learn as much or more about human social behavior by studying social carnivores as by studying nonhuman primates. Byrne (1995), who otherwise takes a strongly primatocentric view of animal cognition, observed that we might learn more about the phylogenetic distribution of what he calls intelligence by doing comparative research. Furthermore, Povinelli & Cant (1995) suggest that the performance by arboreal ancestors of the great ape/human clade of 'unusual locomotor solutions . . . drove the evolution of self-conception' (p. 400). Many nonprimate mammals also perform complex, flexible, and unusual acrobatic motor patterns (locomotor-rotational movements) during social play, and it would be premature to rule out the possibility that the performance of these behaviors is also important to the evolution of self-conception in nonprimates. In some instances it is difficult to know whether aboreal clambering or the performance of various acrobatic movements during play may more be related to the evolution of (mere) body awareness (e.g. knowing one's place in space) and not a concept of self.

What is play?

As other papers in this volume show, the term 'play' covers a wide range of behavior patterns. In this respect it is not different from terms such as 'feeding' or 'mating', both of which may encompass a variety of quite different behaviors when comparing members of either the same or different species. However, unlike play, feeding and mating correspond to easily identified biological functions.

Play is not easily defined (Bekoff & Byers 1981; Fagen 1981; Martin & Caro 1985; Burghardt 1998). Attempts to define it functionally face the problem that it is not obvious that play serves any particular function either at the time at which it is performed or later in life. Indeed several authors have been tempted into defining play as functionless behavior. Alternatively it has sometimes been suggested that play serves some general functions such as improving the motor and cognitive skills of young animals, yielding possible payoffs, for example, in the hunting, foraging, or social abilities of these animals from the time of the play throughout their entire lifespans. Even if this is correct, the reproductive fitness consequences of play may typically be so far removed in the life time of the organisms involved that it would be very difficult to collect data to support the assertion that play increases fitness. Furthermore, there may be different possible evolved functions of play depending on the species being studied. It is difficult to design experiments to test hypotheses about functions of play that are both practicable and ethical. Thus play seems to be either functionless or it can be considered as serving different functions for individuals of different species, ages, and sex (Bekoff & Byers 1981; Fagen 1981; Byers & Walker 1995; Burghardt 1996; Watson & Croft 1996).

These considerations led Bekoff & Byers (1981, pp. 300–1; see also Martin & Caro 1985) to eschew a functional characterization of play by offering the following definition: '*Play* is all motor activity performed postnatally that *appears* [our emphasis] to be purposeless, in which motor patterns from other contexts may often be used in modified forms and altered temporal sequencing. If the activity is directed toward another living being it is called *social play*.' This definition centers on the structure of play sequences – what animals do when they play – and not on possible functions of play. Nonetheless the definition is not without problems, for it would seem to apply, for example, to stereotypical behaviors such as the repetitive pacing or excessive self-grooming sometimes evinced by caged animals. It is difficult to see how to state a non-arbitrary restriction on the range of behaviors that may constitute play.

Because it is not easily defined, play, both social and nonsocial, has been a very difficult behavioral phenotype with which to deal rigorously. A few people would claim that only humans engage in play, but most agree that nonhumans play despite finding it difficult to offer an exceptionless definition. But this lack of a comprehensive definition need not be an impediment to conducting solid research. Our view is that the study of play ought to be approached like the study of any other (putative) natural kind of behavior (Allen & Bekoff 1994). To study play, one ought to start with examples of behaviors which superficially appear to form a single category – those that would be initially agreed upon as play – and look for similarities among these examples. If similarities are found, *then* we can ask whether they provide a basis for useful generalizations. We therefore propose to proceed on the basis of an intuitive understanding of play, guided to some extent by Bekoff & Byers' attempt to define it, but without the view that this or any other currently available definition strictly includes or excludes any specific behaviors from the category of play.

Can there be an evolutionary biology of play? The possible problem of intentionality

Alexander Rosenberg (1990) presents some challenges to an evolutionary approach to social play. One of his concerns hinges on his claim that play is an intentional activity. For reasons similar to those of Dennett (1969) and Stich (1983), and rejected by Allen & Bekoff (1994, 1997), Rosenberg believes that intentional explanations are not suitable for scientific explanations of behavior. Rosenberg, for instance, suggests that it might be inappropriate to attribute the concept of mouse-catching to a cat by asking 'Does it have the concept of mouse, *Mus musculus* in Linnaean terms?' (p. 184). Our view is that possession of the Linnaean concept of a mouse is not a reasonable requirement to be placed on the attribution of beliefs about mouse catching (see Allen & Bekoff 1994).

Rosenberg also argues that there can be no unified evolutionary account of play because actual cases of play have heterogeneous causes and effects, and different underlying mechanisms. He draws an analogy between play and clocks, pointing out that because there are so many different mechanisms that constitute clocks there is no 'single general explanatory theory that really explains what clocks do, how and why they do it.' (p. 180) The problem with this argument is that the kind of 'single general explanatory theory' referred to is not (and should not be) the kind of thing evolutionary biology is necessarily concerned with. While it is the concern of some branches of biology (particularly molecular and cellular) to explain *how* certain organs do what they do, other branches of evolutionary theory are concerned with *what* they do and *why* they do it. So while it would be foolish to expect a singular molecular or cellular account of light-sensing capabilities across species, it is not foolish to expect unity in some aspects of the evolutionary explanations of the development of such organs (although, of course, there will be differences in the evolutionary histories across different species). If Rosenberg was right, there could be no general evolutionary theory of predation or sexual selection by mate choice, for these phenomena too depend on a very heterogeneous set of mechanisms. Play, we submit, is in no worse shape than these well entrenched targets of biological explanation.

Play, pretense, and intentionality

> After all, from an evolutionary point of view, there ought to be a high premium on the veridicality of cognitive processes. The perceiving, thinking organism ought, as far as possible, to get things right. Yet pretense flies in the face of this fundamental principle. In pretense we deliberately distort reality. How odd then that this ability is not the sober culmination of intellectual development but instead makes its appearance playfully and precociously at the very beginning of childhood.
> *(Leslie 1987, p. 412)*

As we noted above, discussions of play commonly refer to the concept of pretense. Because pretense seems to be a fairly sophisticated cognitive ability it has led some authors to deny that nonhuman animals can be said to engage in play. Rosenberg (1990), for example, associates pretense with 'third-order' intentionality (Grice 1957; Dennett 1983, 1987). According to Rosenberg, for animal *a* truly to be playing with *b*, it must be that '*a* does *d* [the playful act] with the intention of *b*'s recognizing that *a* is doing *d* not seriously but playfully. So, *a* wants *b* to believe that *a* wants to do *d* not seriously but with other goals or aims.' (Rosenberg 1990, p. 184) This is third-order because there are three levels of mental state attribution involved, i.e. *a believes* that *b believes* something about *a* 's *desires*. This requirement might be thought to rule out play not just in nonhuman animals, but also in human infants.

In contrast to this approach, the Bekoff & Byers characterization of play is neutral about the intentionality of play behavior. Ultimately it might be found that play is an intentional activity but it would be premature, in our view, to include this in the definition of play. The relevance of intentionality to play is a matter for empirical investigation, and any empirical investigation of the connection between play and intentionality will be shaped by the account of intentionality that is provided (for discussion see Allen & Bekoff 1997, Chapter 6).

From Dennett's intentional stance, organisms are modeled as representing various aspects of their environments and their actions are guided by those representations. For some organisms, these representations may themselves contain information about how other organisms represent their environments. Such a representation of a representation is a case of second-order intentionality in Dennett's scheme. Dennett treats higher-order intentionality as cognitively more sophisticated (and therefore more recently evolved) than first-order intentionality (which in turn is more sophisticated than zero-order or non-intentionality). Thus, to place cognitive capacities into an evolutionary framework, he thinks it is important to identify the distribution of higher-order intentionality among animals.

Millikan (1984) provides a contrasting approach to intentionality. According to Millikan's account, intentionality is a *functional* property – attributions of intentionality provide information about the historical role of a particular trait but do not directly explain or predict the operations of that trait. To understand this it is useful to consider a non-intentional example of a functional property: the function of a sperm to penetrate an egg. Even knowing this function, one cannot predict that any particular sperm will penetrate an egg – it is far more likely that it will not. Likewise, in intentional cases, one cannot predict that any particular organism will act in a way that is rationally predicted by attributing a state with intentional content. While it may be a function of that intentional state to produce the behavior, there is no more guarantee that a state such as a belief or a desire will fulfill its function than there is that a sperm will penetrate an egg. (See Bekoff & Allen 1992 for a discussion of why Millikan's theory is useful for informing and motivating studies in cognitive ethology.)

Different theories of intentionality have different consequences for specifying the contents of intentional states. Consider Dennett's intentional stance first. To attribute a belief in the conjunction of *P* and *Q* entails the attribution of the belief that *Q* for it would be irrational to fail to infer *Q* from the conjunction. Attributing this (rather minimal) rationality to subjects thus seems to entail that any subject capable of believing a conjunction must also be capable of believing each conjunct separately. But in Millikan's framework it is quite possible to have an intentional icon whose function it is to map onto the conjunction of *P* and *Q* *without* the system having either the ability or the tendency to represent the singular *Q*. Imagine, for example, a system whose *Q*-detector only becomes operative once its *P*-detector registers an occurrence of *P*. Such a system would be capable of representing the conjunction of *P* and *Q* without being able to represent *Q* alone. Perhaps, because *Q* rarely occurs in isolation, or when it does its occurrence is normally irrelevant to the organism, it was never important for the members of the species to have evolved isolated *Q*-detectors or the capacity for representing *Q* alone.

Our point at present is not to adjudicate between these different conceptions of intentionality. Rather, each provides a framework within which one may ask different kinds of questions about the behavior of animals. As such, each provides opportunities for research. Dennett's framework emphasizes orders of intentionality as a significant evolutionary variable, and Dennett (1983) suggests experiments that one might perform with vervet monkeys (*Cercopithecus aethiops*) to test his ideas. Dennett is also concerned to explain how animals may sometimes show evidence of higher-order capabilities while at other times or in other contexts showing a lack of ability to reason at a similarly high level – a phenomenon that would be puzzling if the animals were ideally rational. But from within Millikan's perspective this puzzle does not arise. This is because intentional states which are supposed (evolutionarily) to correspond to the intentional states of other organisms (second-order content) need not be related by inference to any general ability to form states with second-order intentional content. An animal may have very specific cognitive abilities with respect to particular intentional states of other organisms, without having the general ability to attribute intentional states to those organisms.

Returning to Rosenberg's third-order account of pretense we see that whether or not one regards it as plausibly attributed to nonhuman animals depends on the general account of intentionality that is adopted. From the intentional stance, if *a* believes that *b* believes that *a* desires to play (third-order) it would seem that ideal rationality would also require that *a* believes that *b* has a belief (second-order). But from a Millikanian perspective this more general second-order belief, if it requires *a* to have a general belief detector, may actually be more sophisticated than the third-order belief which supposedly entails it. A general belief detector may be much more difficult to evolve than a specific belief detector, for the detection of specific beliefs may be accomplished by the detection of correspondingly specific cues.

If this is correct, then on Millikan's account Jethro (Marc's dog) may be capable of the third-order belief that (or, at least, a state with the intentional content that) Sukie (Jethro's favorite canid play pal) wants Jethro to believe that her bite was playful and not aggressive, even though Jethro is perhaps limited in his ability to represent and hence think about Sukie's second-order desires in general. Further below we shall argue for such an understanding of the content of play signals using Millikan's approach to intentionality.

If one takes a Dennettian approach to third-order intentionality, then Rosenberg's third-order analysis of pretense seems over-inflated. It is doubtful that many animals could make the general inferences that the rationality assumption seems to require them to be capable of making from any specific third-order belief. A particular behavioral sequence in social play may involve pretense even though neither participant has a general conception of pretense. In social play, an animal, *a*, may, for example, bare her teeth in a gesture that might also occur during or as a prelude to a fight. The playmate, *b*, may respond by growling – another behavior that could occur during a fight. The first animal, *a*, may then pounce on *b* and grasp some portion of *b*'s body between her teeth. This sequence involves motor patterns found in fighting, yet the animals are not fighting. What cognitive abilities must *a* and *b* possess for this to be possible? They must be capable of discriminating those occasions when a behavior is genuinely aggressive from those when it is playful. This could be achieved by detecting subtle differences between, for example, aggressive teeth baring and playful teeth baring – if such differences exist. In the only study of its type of which we are aware, Hill & Bekoff (1977) found that bites directed towards the tail, flank, legs, abdomen, or back lasted a significantly shorter time and were more stereotyped during social play than during aggression in Eastern coyotes. Or it can be achieved by providing contextual cues that inform players about the difference between aggression and play. As we shall discuss below, in many species signals have evolved to support the second approach, and such signals may be understood as intentional icons that convey the messages about the intentions of the play participants.

Play signals

When animals play they typically use action patterns that are also used in other contexts, such as predatory behavior, antipredatory behavior, and mating. These action patterns may not be intrinsically different across different contexts, or they may be hard to discriminate even for the participants. To solve the problems that might be caused by, for example, confusing play for mating or fighting, many species have evolved signals that function to establish and maintain a play 'mood' or context. In most species in which play has been described, play-soliciting signals appear to foster some sort of cooperation between players so that each responds to the other in a way consistent with play and different from the responses the same actions would elicit in other contexts (Bekoff 1975, 1978, 1995b; Bekoff & Byers 1981; Fagen 1981; Bekoff & Allen 1992). Play-soliciting signals also provide aid to the interpretation of other signals by the receiver (Hailman 1977, p. 266). Coyotes, for example, respond differently to threat gestures in the absence of any preceding play signal than they do to threat gestures that are immediately preceded by a play signal or in the middle of sequence that was preceded by a play signal (Bekoff 1975). Given the possible risks that are attendant on mistaking play for another form of activity, it is hardly surprising that animals should have evolved clear and unambiguous signals to solicit and maintain play.

The canid 'play bow', a highly ritualized and stereotyped movement that seems to function to stimulate recipients to engage (or to continue to engage) in social play (Bekoff 1977), provides an excellent example of what we are calling a play signal and it has been extensively studied in this context. That play bows are important for initiating play is illustrated by the example of a dominant female coyote pup who was successful in initiating chase play with her subordinate brother on only 1 of 40 (2.5%) occasions. Her lone success occurred on the only occasion in which she had signaled previously with a bow, although on the other occasions she engaged in a variety of behaviors that are sometimes successful in initiating play such as rapid approach/withdrawals, exaggerated pawing toward the sibling's face, and head waving and low grunting (Bekoff, 1975).

To say that the bow is stereotyped is to say that the form that play bows take is highly uniform without implying anything about the contextual versatility with which bows are used. When performing a bow, an individual crouches on its forelimbs, remains standing on its hindlegs, and may wag its tail and bark. The bow is a stable posture from which the animal can move easily in many directions, allowing the individual to stretch its muscles before and while engaging in play, and places the head of the bower below another animal in a non-threatening position. Play-soliciting signals show little (but some) variability in form or temporal characteristics (Bekoff 1977). The first play bows that very young canids have been observed to perform are highly stereotyped, and learning seems to be relatively unimportant in their development. The stereotyped nature of the play bow is probably important for avoiding ambiguity.

Play bows occur throughout play sequences, but most commonly at the beginning or towards the middle of playful encounters. In a detailed analysis of the form and duration of play bows (Bekoff 1977) it was shown that duration was more variable than form, and that play bows were always less variable when performed at the beginning, rather than in the middle of, ongoing play sequences. Three possible explanations for this change in variability include (1) fatigue, (2) the fact that animals are performing them from a wide variety of preceding postures, and (3) there is less of a need to communicate that this is still play than there is when trying to initiate a new interaction. These explanations are not exclusive alternatives.

The meaning of play bows

Play bows occur almost exclusively in the context of play, and it is common to gloss play-soliciting signals with the message 'what follows is play' or 'this is still play'. What is the significance of these glosses for the players themselves? Are they in any way aware of the meaning of the play bows, or are they simply conditioned to respond differently, e.g. less aggressively or less sexually, when a specific action such as a bite or a mount is preceded by a play bow?

One way to approach this question is to ask whether play signals such as bows are used to maintain social play in situations where the performance of a specific behavior during a play bout could be misinterpreted. A recent study of the structure of play sequences (Bekoff 1995b) showed that bows in some canids, infant and adult domestic dogs, infant coyotes, and infant wolves, are often used immediately before and after an action that can be misinterpreted and disrupt ongoing social play. Recall that the social play of canids (and of other mammals) contains actions, primarily bites, that are used in other contexts that do not contain bows (e.g. agonistic, predatory, or antipredatory). Actions such as biting accompanied by rapid side-to-side shaking of the head are used in aggressive interactions and also during predation and could be misinterpreted when used in play.

Bekoff asked the following questions: (1) What proportion of bites directed to the head, neck, or body of a play partner and accompanied by rapid side-to-side shaking of the head are immediately preceded or followed by a bow? (2) What proportion of behavior patterns other than bites accompanied by rapid side-to-side shaking of the head are immediately preceded or followed by a bow? Actions considered here were mouthing or gentle biting during which the mouth is not closed tightly and rapid side-to-side shaking of the head is not performed, biting without rapid side-to-side shaking of the head, chin-resting, mounting from behind (as in sexual encounters), hip-slamming, standing-over assertively, incomplete standing-over, and vocalizing aggressively (for descriptions see Bekoff 1974; Hill & Bekoff 1977). Not considered was the situation in which the recipient of bites accompanied by rapid side-to-side shaking of the head performed a bow immediately before or immediately after its partner performed bites accompanied by rapid side-to-side shaking of the head or other actions, because these rarely occurred. It was hypothesized that if bites accompanied by rapid side-to-side shaking of the head or other behavior patterns could be or were misread by the recipient, resulting in a fight, for example, then the animal who performed such actions might have to communicate that they were performed in the context of play and were not meant to be taken as an aggressive or predatory move. On this view, bows would not occur randomly in play sequences; the play atmosphere would be reinforced and maintained by performing bows immediately before or after actions that could be misinterpreted.

The results of Bekoff's study support the inference that bows might serve to provide information about other actions that follow or precede them. In addition to sending the message 'I want to play' when they are performed at the beginning of play, bows performed in a different context, namely during social play, might also carry the message 'I want to play despite what I am going to do or just did – I still want to play' when there might be a problem in the sharing of this information between the interacting animals. Species differences were also found that can be interpreted by what is known about variations in the early social development of these canids (Bekoff 1974; see also Feddersen-Petersen 1991). The interspecific differences also are related to the question at hand. For example, infant coyotes are much more aggressive and engage in significantly more rank-related dominance fights than either the infant (or adult) dogs or the infant wolves who were studied. During the course of this study, no consistent dominance relations were established in either the dogs or the wolves, and there were no large individual differences among the play patterns that were analyzed in this study. Social play in coyotes typically is observed only after dominance relationships have been established in paired interactions. Coyotes appear to need to make a greater attempt to maintain a play atmosphere, and indeed, they seem also to need to communicate their intentions to play *before* play begins more clearly than do either dogs or wolves who have been studied (Bekoff 1975, 1977). Subordinate coyote infants are more solicitous and perform more play signals later in play bouts. These data suggest that bows are not non-randomly repeated merely when individuals want to increase their range of movement or stretch their muscles. However, because, among other things, the head of the bowing individual is usually below that of the recipient, bowing may place the individual in a non-threatening, self-handicapping, posture. Self-handicapping might occur when the bowing animal is dominant or subordinate to her partner: when the bower is dominant she may be sending the message 'I do not want to dominate you more' and when the bower is subordinate she may be sending the message 'I am not trying to dominate you.'

Standing-over, which usually is an assertion of dominance in infant coyotes (Bekoff, 1974) but not in infant beagles or wolves of the same age was followed by a significantly higher proportion of bows in coyotes when compared to dogs or to infant wolves. Because bows embedded within play sequences were followed significantly more by playing than by fighting after actions that could be misinterpreted were performed (unpublished data), it does not seem likely that bows allow coyotes (or other canids) more readily to engage in combat, rather than play, by increasing their range of movement, although this possibility can not presently be ruled out in specific instances.

In addition to the use of signals such as bows, it is also possible that the greater variability of play sequences when compared to sequences of agonistic behavior (Bekoff & Byers 1981) allows animals to use the more varied sequences of play as a composite play signal that helps to maintain the play mood; not only do bows have signal value but so also do play sequences (Bekoff 1976; 1977). Self-handicapping occurring, for example, when a dominant individual allows itself to be dominated by a subordinate animal, also might be important in maintaining on-going social play (Altmann 1962). Watson & Croft (1996) found that red-neck wallabies (*Macropus rufogriseus banksianus*) adjusted their play to the age of their partner. When a partner was younger, the older animal adopted a defensive, flat-footed posture, and pawing rather than sparring occurred. In addition, the older player was more tolerant of its partner's tactics and took the initiative in prolonging interactions. While more data are needed, this study also suggests that the benefits of play may vary according to the age of the player.

Putting play in a broader cognitive context

The data presented above suggest that at least some canids (and most likely other mammals) cooperate when they engage in social play, and may negotiate these cooperative ventures by sharing their intentions. Fagen (1993, p. 192) has also noted that 'Levels of cooperation in play of juvenile primates may exceed those predicted by simple evolutionary arguments . . . ' In general, animals engaged in social play use specific signals to modulate the effects of behavior patterns that are typically performed in other contexts, but whose meaning is changed in the context of play. These signals are often flexibly related to the occurrence of events in a play sequence that might violate expectations within that sequence. Furthermore, the relationship of play to a cognitive appreciation of the distinction between reality and pretense provides an important link to other cognitive abilities, such as the ability to detect deception or to detect sensory error. Given these connections, a detailed consideration of some selected aspects of social play might help promote the development of more sophisticated theories of consciousness, intentionality, representation, and communication.

The ability to engage in pretend play (e.g., to manipulate an object as if it is something else) normally first appears in human children around 12 months of age (Flavell et al. 1987). This is well before children appear to be capable of attributing mental states to others. Human children also seem capable of engaging in social play before they have a developed theory of mind. Leslie, in the quotation given earlier, expresses surprise about the distortion of reality implied by pretense. We, however, are inclined to suggest that play is one way that an animal may learn to discriminate between its perceptions of a given situation and reality, learning, for example to differentiate a true threat from a pretend threat. From this perspective it would be perhaps more surprising if cognitively sophisticated creatures could get to this point without the experiences afforded by play (for related discussion see Parker & Milbraith 1994).

It is also possible that experiences with play promote learning about the intentions of others. Even if the general capacity for understanding the mental states of others is a specifically human trait, many other species may be able to share information about particular intentions, desires, and beliefs. How might a play bow serve to provide information to its recipient about the sender's intentional state? It is possible that the recipient shares the intentions (beliefs, desires) of the sender based on the recipient's own prior experiences of situations in which she performed bows. Given our earlier discussion of specialized mechanisms, it may be reasonable to attribute a very specific second-order inference of the form 'when I bow I want to play so when you bow you want also to play' without being committed to a general capacity for the possession of second-order mental states in these animals.

Recently, Gopnik (1993, p. 275) has argued that ' . . . certain kinds of information that comes, literally, from inside ourselves is coded in the same way as information that comes observing the behavior of others. There is a fundamental cross-modal representational system that connects self and other.' Gopnik (see also Meltzoff & Gopnik 1993) claims that others' body movements are mapped onto one's own kinesthetic sensations, based on prior experience of the observer, and she supports her claims with discussions of imitation in human newborns.

For example, Gopnik wants to know if there is an equivalence between the acts that infants see others do and the acts they perform themselves, and imagines 'that there is a very primitive and foundational "body scheme" that allows the infant to unify the seen acts of others and their own felt acts into one framework' (Gopnik 1993, p. 276). If by 'primitive and foundational' Gopnik means phylogenetically old, then there should be some examples, or at least precursors, of this ability in other animals. Gopnik and her colleague Andrew Meltzoff also consider the possibility that there is 'an innate mapping from certain kinds of perceptions of our own internal states . . . In particular, we innately map the body movements of others onto our own kinesthetic sensations. This initial bridge between the inside and the outside, the self and other, underlies our later conviction that all mental states are things both we and others share' (Gopnik 1993, p. 275; see also Flanagan 1992, pp. 102ff).

How these ideas might apply to nonhuman animals awaits further study. There are preliminary suggestions that Gopnik's ideas might enjoy some support from comparative research on animal cognition. For example, Savage-Rumbaugh (1990, p. 59) noted that 'Likewise, if Sherman screams when he is upset or hurt, Sherman may deduce that Austin is experiencing similar feelings when he hears Austin screams.

This view is supported by the observation that Sherman, upon hearing Austin scream, does not just react, but searches for the cause of Austin's distress.' This cause–effect relationship is generated after sufficient experience – if an animal screams when he is upset or hurt he may deduce that another is experiencing similar feelings when he hears a scream. Tomasello et al. (1989) also note that some gestures in chimpanzees may be learned by 'second-person imitation' – 'an individual copying a behavior directed to it by another individual' (p. 35). They conclude (p. 45) that chimpanzees ' . . . rely on the sophisticated powers of social cognition they employ in determining what is perceived by a conspecific and how that conspecific is likely to react to various types of information . . . '

There is also the possibility that in social play one dog might be able to know that another dog wants to play by knowing what she feels like when she performs a play bow. Among the questions that need to be studied in detail is, 'Does a dog have to have performed a bow (or other action) before knowing what a bow means and subsequently being able to make attributions of mental states to other individuals?' The following two hypotheses would have to be distinguished: (1) viewing a play bow induces a play mood in the recipient because of kinesthetic mapping and (2) viewing a play bow induces knowledge in the recipient of how the actor feels. With respect to bows, at least, there are data that suggest that there is a genetic component to them; the first bows that are observed to be performed by young canids are highly stereotyped and occur in the correct social context (Bekoff 1977). Could these data support Gopnik's idea about the 'primitive and foundational "body scheme"? And, if so, how is learning incorporated into the development of social communication skills? Regardless of how nature and nurture mix, sparse evidence at hand supports the view that studies of animal cognition can inform the study of human cognition, and that much more comparative research is needed.

Concluding remarks: social play and comparative studies of animal cognition

Because social play is a widespread phenomenon, especially among mammals, it offers the opportunity for much more truly comparative and evolutionary work on intentionality, communication, and information sharing (see also Parker & Milbraith 1994). The collection of new data will provide for a much broader perspective on the origins of intentionality in diverse species. Nonetheless, some primatologists write as if theirs are the only subjects who are capable of recognizing the intentions of others. For example, Byrne (1995, p. 146) writes: '...great apes are certainly "special" in some way to do with mentally representing the minds of others. It seems that the great apes, especially the common chimpanzee, can attribute mental states to other individuals; but no other group of animals can do so – apart from ourselves, and perhaps cetaceans.' To dismiss the possibility that nonprimates are capable of having a theory of mind, not only do more data need to be collected, but existing data about intentionality in nonprimates need to be reconsidered (see also Beck 1982 on chimpocentrism). Furthermore, claims about the uniqueness of nonhuman primates are often based on very few comparative data derived from tests on small numbers of nonhuman primates who might not be entirely representative of their species. The range of tests that have been used to obtain evidence of intentional attributions is also extremely small, and such tests are often biased towards activities that may favor apes over monkeys or the members of other species. There is evidence (Whiten & Ham 1992) that mice can outperform apes on some imitation tasks. These data do not make mice 'special'; rather they suggest that it is important to investigate the abilities of various organisms in respect to their normal living conditions. The study of social play affords this opportunity.

Acknowledgments

Some of the material in this chapter has been excerpted from Allen & Bekoff (1997) with permission of The MIT Press. Colin Allen was supported by NSF grant SBR-9320214 during the writing of this chapter. Maxeen Biben provided helpful comments.

References

Allen, C. & Bekoff, M. 1994. Intentionality, social play, and definition. *Bio. Phil.*, **9**, 63–74.

Allen, C. & Bekoff, M. 1997. *Species of Mind: The Philosophy and Biology of Cognitive Ethology*. Cambridge, Massachusetts: MIT Press.

Altmann, S. A. 1962. Social behavior of anthropoid primates: Analysis of recent concepts. In: *Roots of Behavior* (eds. by E. L. Bliss), pp. 277–85. New York: Harper.

Beck, B. B. 1982. Chimpocentrism: bias in cognitive ethology. *J. Human Evol.*, **11**, 3–17.

Bekoff, M. 1974. Social play and play-soliciting by infant canids. *Amer. Zool.*, **14**, 323–40.
Bekoff, M. 1975. The communication of play intention: Are play signals functional? *Semiotica*, **15**, 231–9.
Bekoff, M. 1976. Animal play, problems and perspectives. *Persp. Ethol.*, **2**, 165–88.
Bekoff, M. 1977. Social communication in canids, Evidence for the evolution of a stereotyped mammalian display. *Science*, **197**, 1097–9.
Bekoff, M. 1978. Social play, Structure, function, and the evolution of a cooperative social behavior. In: *The Development of Behavior: Comparative and Evolutionary Aspects* (eds. G Burghardt & M. Bekoff), pp. 367–83. New York: Garland.
Bekoff, M. 1995a. Cognitive ethology and the explanation of nonhuman animal behavior. In: *Comparative Approaches to Cognitive Science* (eds. H. L. Roitblat and J.-A. Meyer), pp. 119–50. Cambridge: Massachusetts: MIT Press.
Bekoff, M. 1995b. Play signals as punctuation, The structure of social play in canids. *Behaviour*, **132**, 419–29.
Bekoff, M. 1998. Playing with play, What can we learn about evolution and cognition? In: *The Evolution of Mind* (eds. D. Cummins & C. Allen) New York: Oxford University Press.
Bekoff, M. & Allen, C. 1992. Intentional icons: towards an evolutionary cognitive ethology. *Ethology*, **91**, 1–16.
Bekoff, M. & Byers, J. A. 1981. A critical reanalysis of the ontogeny of mammalian social and locomotor play, An ethological hornet's nest. In: *Behavioral Development, The Bielefeld Interdisciplinary Project* (eds. K. Immelmann, G. W. Barlow, L. Petrinovich, and M. Main), pp. 296–337. New York: Cambridge University. Press.
Bekoff, M. & Jamieson, D. (ed.) (1996). *Readings in Animal Cognition*. Cambridge, Massachusetts: MIT Press.
Burghardt, G. M. 1998. Play. In: *Encyclopedia of Comparative Psychology* (eds. G. Greenberg and M. Haraway). New York: Garland.
Byers, J. A. & Walker, C. 1995. Refining the motor training hypothesis for the evolution of play. *Amer. Nat.*, **146**, 25–40.
Byrne, R. 1995. *The Thinking ape: Evolutionary Origins of Intelligence*. New York: Oxford University Press, .
Darwin, C. 1871/1936. *The Descent of Man and Selection in Relation to Sex*. New York: Random House; Modern Library edition.
Dennett. D. C. 1969. *Content and consciousness*. New York: Routledge and Kegan Paul.
Dennett, D. C. 1983 Intentional systems in cognitive ethology, The 'Panglossian paradigm' defended. *Behav. Brain Sci.*, **6**, 343–90.
Dennett, D. C. 1987. *The Intentional Stance*. Cambridge, Massachusetts: MIT Press.
Fagen, R. M. 1981. *Animal Play Behavior*. New York: Oxford University Press.
Fagen, R. 1993. Primate juveniles and primate play. In: *Juvenile Primates: Life History, Development, and Behavior* (eds. M. E. Pereira and L. A. Fairbanks), pp. 183–96. New York: Oxford University Press.
Feddersen-Petersen, D. 1991. The ontogeny of social play and agonistic behaviour in selected canid species. *Bonn. Zool. Beitr.*, **42**, 97–114.
Flanagan, O. J. 1992. *Consciousness Reconsidered*. Cambridge, Massachusetts: MIT Press.

Flavell, J., Flavell, E, & Green, F. 1987. Young children's knowledge about the apparent–real and pretend–real distinctions. *Dev. Psych.*, **23**, 816–22.

Gopnik, A. 1993. Psychopsychology. *Consciousness and Cognition*, **2**, 264–80.

Grice, H. P. 1957. Meaning. *Phil. Rev.*, **66**, 377–88.

Hailman, J. P. 1977. *Optical Signals: Animal Communication and Light.* Bloomington, Indiana: Indiana University Press.

Hill, H. L. & Bekoff, M. 1977. The variability of some motor components of social play and agonistic behaviour in infant eastern coyotes *Canis latrans var. Anim. Behav.*, **25**, 907–9.

Huber, P. 1810. *Recherche sur les moeurs des fourmis indigènes.* Paris, Geneve: J. J. Paschoud.

Jamieson, D. & Bekoff, M. 1993. On aims and methods of cognitive ethology. *Phil. Sci. Assoc.*, **2**, 110–24.

Leslie, A. M. 1987. Pretense and representation. The origins of 'theory of mind.' *Psych. Rev.*, **94**, 412–26.

Martin, P. & Caro. T. M. 1985. On the functions of play and its role in behavioral development. *Adv. in the Study of Behavior*, **15**, 59–103.

Meltzoff, A., & Gopnik, A. 1993. The role of imitation in understanding persons and developing a theory of mind. In: *Understanding Other Minds* (eds. S. Baron-Cohen, H. Tager-Flusberg, and D. Cohen), pp. 335–66. New York: Oxford University Press.

Millikan, R. G. 1984. *Language, Thought, and Other Biological Categories.* Cambridge, Massachusetts: MIT Press.

Parker, S. T. & Milbraith, C. 1994. Contributions of imitation and role-playing games to the construction of self in primates. In *Self-Awareness in Animals and Humans: Developmental Perspectives* (eds. S. T. Parker, R. W. Mitchell, and M.L. Boccia), pp. 108–128. New York: Cambridge University Press.

Povinelli, D. J. & Cant, J. G. H. 1995. Arboreal clambering and the evolution of self-conception. *Q. Rev. Bio.*, **70**, 393–421.

Rosenberg, A. 1990. Is there an evolutionary biology of play? In: *Interpretation and Explanation in the Study of Animal Behavior, Vol. 1, Interpretation, Intentionality, and Communication* (eds. M. Bekoff & D. Jamieson), pp. 180–96. Boulder, Colorado: Westview Press. (Reprinted in Bekoff & Jamieson 1996.)

Savage-Rumbaugh, E. S. 1990. Language as a cause-effect communication system. *Phil. Psych.*, **3**, 55–76.

Schaller, G. B. & Lowther, G. R. 1969. The relevance of social carnivore behavior to the study of early hominids. *Southwest J. Anthro.*, **25**, 307–41.

Stich, S. 1983. *From Folk Psychology to Cognitive Science.* Cambridge, Massachusetts: MIT Press.

Tinbergen. N. 1972. Foreword to Hans Kruuk. *The Spotted Hyena.* Chicago: University of Chicago Press.

Tomasello, M., Gust, D., & Frost, G. T. 1989. A longitudinal investigation of gestural communication in young chimpanzees. *Primates*, **30**, 35–50.

Watson, D. M. & Croft, D. B. 1996. Age-related differences in playfighting strategies of captive male red-necked wallabies (*Macropus rufogriseus banksianus*). *Ethology*, **102**, 336–46.

Whiten, A. & Ham, R. 1992. On the nature and evolution of imitation in the animal kingdom: Reappraisal of a century of research. *Adv. Study Behav.*, **21**, 239–83.

6

The structure–function interface in the analysis of play fighting

SERGIO M. PELLIS and VIVIEN C. PELLIS
Department of Psychology and Neuroscience, University of Lethbridge, Lethbridge, Alberta, Canada T1K 3M4

Introduction

In the ongoing commentary on Smith's (1982) target article on play, Moran (1985) wrote a critique entitled 'Behavioral description and its impact on functional inference.' In response, Smith (1985) wrote a counter-critique entitled 'Functional hypotheses and their impact on behavioral description.' While seemingly at odds, these two points of view are actually complementary. If play fighting did not resemble serious fighting, in at least some crude ways, then it is unlikely that the hypothesis that the former served as practice for the latter (Groos 1898) would ever have been considered. In this regard, description preceded functional inference. However, once formulated, such an hypothesis makes predictions about other features of the behavior, in this case, play fighting, which are not known. Therefore, behavioral description informs functional inference, which in turn, influences further description.

That play fighting functions as a means of refining the skills necessary for combat is an hypothesis which continues to receive widespread support (e.g., Caro 1988; Fagen 1981; Pellis 1981b; Smith 1982; Symons 1978a). Indeed, the hypothesis that play serves as practice is widely endorsed in lay treatments of the topic, as can be ascertained from both television nature documentaries and from popular books. For example, Angier (1995) asserts that 'through play, animals can rehearse many of the moves they will need as adults' (p. 133). With regard to play fighting, this hypothesis asserts that this form of play serves to practice the tactics of attack and defense, which would otherwise only occur in dangerous situations (Symons 1978a). The enhancement of fighting skills that such practice affords is often argued to be responsible for shaping the form of play fighting (Smith 1982). It should also be noted that for current purposes, the play fighting of juveniles will be primarily discussed. Play fighting amongst adults, who play less generally (Fagen 1981), may have alternative functions (Rhine 1973). For example, subordinate adult male rats, *Rattus norvegicus,* use playful contact as a means of 'friendship maintenance' with dominants (Pellis et al. 1993), and adolescent chimpanzees, *Pan troglodytes* use play fighting for dominance testing (Paquette 1994).

The practice hypothesis is generally applied to childhood, where play fighting as juveniles is viewed as enhancing combat skills in adulthood. While there is little to no experimental evidence for such skill enhancement (Martin & Caro 1985), many supposed design features of play fighting appear to be consistent with this function (Caro 1988). However, many of these design features take the following form. Males of many species engage in both more play fighting as juveniles and serious fighting as adults than do females. Therefore, heightened play fighting in males is a design feature consistent with the practice of combat related behavior patterns (e.g., Biben 1982; Berger 1980; Byers 1980; Crowell-Davis et al. 1987; Pfeiffer 1985; Symons 1978a; Watson & Croft 1993). Unfortunately, such design features are loosely correlated to combat behavior, and alternative hypotheses are often more parsimonious (e.g., Coppinger & Smith 1989; Pellis et al. 1997). The case for play fighting as having the function of rehearsing adult combat would be more compelling if the structure of play fighting (i.e., the organization of the movement patterns and sequences performed) had design features suitable for this purpose. This chapter will show that the very behavioral features that make play fighting play, also make play fighting a poor means of rehearsing combat skills. While some evidence suggests that such rehearsal may occur for some combat tactics in some situations, the practice hypothesis cannot account for the form of play fighting in most species.

What is play fighting?

Play fighting has been thought to involve the use of species-specific patterns of agonistic behavior in a non-serious context (e.g., Fagen 1981; Meaney et al. 1985). The differences in the form of play fighting between species have been largely attributed to species-specific differences in body targets attacked, in tactics of attack and defense, and in body shape and maneuverability (e.g., Aldis 1975; Symons 1978a). While in many species reported, the targets of attack and defense are the same as those of serious fighting (e.g., Fox 1969; Pellis 1981b; Poole 1966), this does not appear to be the case universally (Pellis 1988). For a variety of muroid rodents, the targets of attack and defense during play fighting are the same as those of pre-copulatory behavior, not aggression (Pellis 1993). For example, male Djungarian hamsters, *Phodopus campbelli* lick and nuzzle the mouth of the female during pre-copulatory contact. Similarly, during play fighting, juveniles of both sexes compete for access to each others' mouths, which are licked and nuzzled if contacted. In contrast, the contact during serious fighting involves biting, and is mostly targeted at the lower flanks and dorsum (Pellis & Pellis 1989). Previous studies of muroid rodents have emphasized the similarity of juvenile play fighting and adult aggression (e.g., Takahashi & Lore 1983; Taylor 1980), and have considered sexual behavior, when it occurred, as interrupting play fighting (Hole & Einon 1984; Poole & Fish 1975). The fact that targets of sexual contact form the organizational basis of play fighting in muroid rodents shows that sexual behavior is not peripheral, but central to the play fighting of these species. The role of sexual targets and behavior patterns in the organization of play fighting may also be underestimated for other species (Pellis 1993). The variability in the content of play fighting needs to be considered with respect to the practice hypothesis.

What is practiced during play fighting?

In a recent paper, Stamps (1995) reviewed the literature on motor learning in humans, and its application to another motor problem common to animals, that of the use of escape routes to foil predators. Importantly, much of this literature involves tasks where the subjects refine motor patterns already in their behavioral repertoire. In reviewing the literature on play and development, Martin & Caro (1985) clearly showed that play is not necessary for the acquisition of species-typical behavior patterns. Rather, motor patterns appear in the repertoire independently of such experience. The issue, they point out, is not whether play is necessary for the genesis of motor patterns, but whether play provides a means of refining the use of those patterns. That is, whether skill in the use of those motor patterns is enhanced by their performance during play. The motor learning literature would then seem highly pertinent to the evaluation of the practice hypothesis as it applies to play fighting. Two of the conclusions from human motor learning are particularly relevant to the evaluation of the suitability of play fighting as a means of providing practice for refining combat skills. Firstly, '... experimental studies suggest that learning one motor program usually has little if any effect on the performance of another motor program. That is, transfer across motor skills is usually low' (Stamps 1995, p. 45). Secondly, '. . .effective learning of serial motor programs may not require that the entire program be practiced in its entirety. Indeed, it may be more efficient to practice the most difficult portions of the sequence, and then string the portions together to form the final motor pattern' (p. 45).

A major problem is that of determining how similar the movement patterns in play fighting have to be to those in serious fighting for the former to provide rehearsal for the latter. In dynamic situations, such as combat, effective motor programs cannot be rigidly stereotyped, but have to be variable, to adapt to changing contexts (Pellis 1985). Nonetheless, specific motor patterns have a basic motoric theme which identifies them as different from each other. For example, one defensive tactic to a bite to the rump in muroid rodents is to turn to face the attacker while continuing to stand on their hindlimbs (Pellis 1989). However, the exact composition of the body movement used to execute this maneuver differs markedly between squat-bodied and elongated-bodied rodents. For hamsters, squat-bodied rodents, the forequarters are raised to about 45° and the body is rotated around its longitudinal axis towards the attacker. In contrast, in voles, which are more elongated in shape than hamsters, the forequarters pivot horizontally around a fixed point at the sacral area. Then, when either type of rodent faces its opponent, the hindquarters move forward and around to align with the new position of the head (Fig. 6.1). While the theme of the motor pattern is the same, there is a species-specific difference in the motor organization used to execute that tactic. Differences in the starting posture of the performer, and the position of the opponent, can greatly modify the kinematics of such a motor pattern, even when executed by the same animal. However, the essential theme remains identifiably constant (Pellis 1985) and hence recognizably different from motor pattterns with other themes.

For play fighting to provide practice for serious fighting, motor patterns used during serious fighting have to be the ones performed. They have to be the same ones in terms of the themes of the action, rather than every execution of the pattern having to be identical. That is, if power punches are to be practiced, it is no use performing jabs. This does not mean that every power punch thrown has to be identical. The kinematics can vary in that the punch can be adapted to the context in which it is used. Therefore, our first expectation from play fighting is that the motor patterns used would be of the same theme as those that are used during serious fighting. A second expectation is that in order for play fighting to be most effective in enhancing combat skills, then those tactics that are

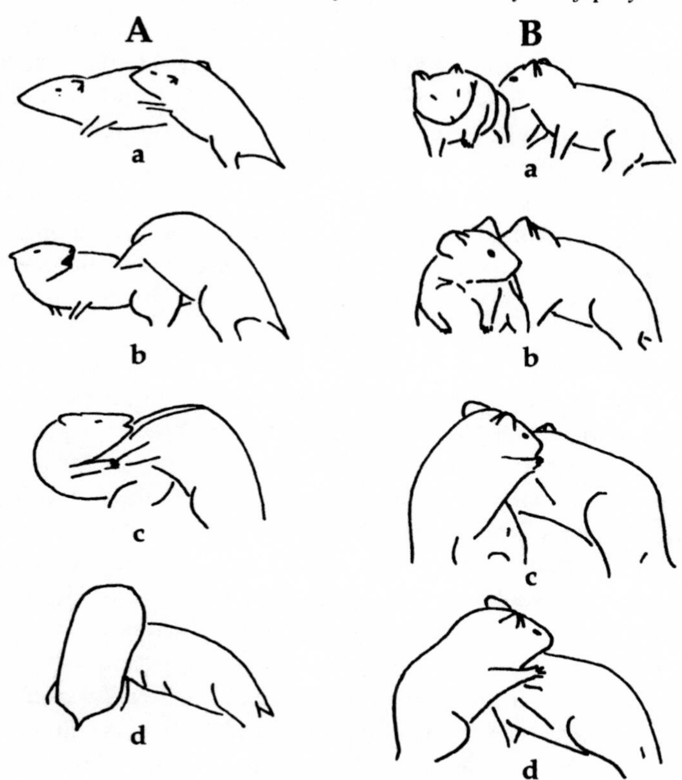

Fig. 6.1 Rump defense by turning laterally toward the attacker is shown for two species of rodents. In A, the attacking vole (on the right side) lungs and bites the defender's lower flank (a) the defender turns its head and neck laterally toward the attacker (b) then, with a lateral turn of its upper body, the defender lunges at the side of the attacker's face (c). Once the defender delivers a retaliatory bite, it pivots around it's hindlimbs and, withdrawing its lower body, rises to an upright posture while simultaneously pressing its attack (d). In B, the attacking hamster (on the right) lungs and bites the defender's rump (a). The defender turns its head and neck laterally towards the attacker (b), but then to complete its turn towards the attacker, the defender raises its forequarters while rotating around its longitudinal axis (c). Once the defender faces the attacker, it pivots around its hindlimbs, withdraws the lower body, and presses the counterattack (d). (From Pellis 1997, © 1997, John Wiley & Sons, Inc).

the most difficult to execute correctly (see below), should receive the most practice. However, our analyses of play fighting suggest that these expectations are rarely met.

The finding that play fighting for some species involves competition over access to sexual but not agonistic targets, is the first piece of evidence from the structure of play fighting that suggests that the practice hypothesis is unlikely to be the universal functional outcome that has shaped the form of play fighting. For example, even though the tactics used during play fighting by rats superficially resemble those used in serious combat (Takahashi & Lore 1983), because different targets are involved, they are markedly modified in form (Pellis & Pellis 1987). During serious fighting, the dorsum and lower flanks are mostly bitten, whereas during play fighting the nape is nuzzled, as may occur during adult pre-copulatory behavior (Pellis 1988). The tactics of attack have little resemblance to those of agonistic attack, whereas those of defense are somewhat similar to those of agonistic defense (Pellis & Pellis 1987). Therefore, play fighting in rats and other muroid rodents provides little opportunity for the performance of many combat-typical tactics. Given that the targets attacked and defended are sexual areas, it is possible that for muroid rodents play fighting is used for the practice of sexual, not aggressive behavior. Indeed, in his original formulation of the practice hypothesis, Groos (1898) noted that play fighting often acts as rehearsal for sex. If so, the practice hypothesis would predict that the tactics rehearsed during play fighting should resemble those occurring during sexual encounters.

Male rats reared in isolation during the juvenile period later exhibit incompetence in sexual performance as adults (Hård & Larsson 1971). Given that rearing juvenile males with non-playful partners still results in these deficits as adults (Einon et al. 1978), it is likely that deprivation of social play rather than social contact in general, is responsible for the deficit in sexual performance in adulthood. In particular, given that play deprived rats have difficulty in achieving the appropriate orientation when mounting (Larsson 1978), it has been suggested that standing on top of a supine play partner, the most commonly occurring juvenile play fighting configuration (Panksepp 1981), provides the opportunity to acquire the necessary orienting skills (Moore 1985). However, this design feature of play fighting in rats is only weakly correlated to inter-animal orientation when mounting. During sexual encounters, females mainly evade males by dodging and running away (McClintock & Adler 1978). Indeed, among wild rats in the field, males often attempt to mount females while chasing them (Whishaw & Whishaw 1996). Therefore, orienting playfully to the nape when the partner is evading would mimic the skills needed during sexual encounters more closely than those that occur during wrestling. If play fighting were used to rehearse correct orientation in mounting, then the use of evasive defensive maneuvers would be more appropriate. Yet in play fighting by juvenile rats it is rotating to supine, not evasion, that is the most frequently used defensive tactic (Pellis & Pellis 1990). Therefore, for rats, and probably for other muroid rodents, the primary function of play fighting does not appear to be to provide practice for either sex or aggression.

The finding that during play fighting animals supposedly use the tactics of attack and defense in the absence of agonistic signals of threat has been held to be strong evidence for the practice hypothesis. This is because play fighting is thought to provide the opportunity for the use of the tactics of attack and defense which would otherwise only be used in a serious context. That is, the tactics of combat, not those of communication, are rehearsed (Symons 1974). For these reasons, it has been suggested that during play fighting, the most difficult tactics (i.e., those of attack and defense) are the ones rehearsed the most, because they will be the most critical ones needed (Symons 1978a). However, even within the range of tactics available, some are more difficult to use than others. Also, it does not appear to be the case that the ones that are more difficult receive the most practice during play fighting. For example, during escalated fights in male deer, head-to-head slamming occurs, which at its most dangerous involves the animals jumping towards their opponents from a distance of up to 2.5 m away (Alvarez 1993). Successful execution involves precise timing and placement of the blows. Errors can lead to injury (Geist 1971). Yet these tactics are not the ones used during play fighting, where gentler head butting and head to head pushing (Alvarez 1993) are the tactics used (Fagen 1981; see Miller & Byers (Chapter 7) for further discussion of ungulate play). Rats provide another example. There are two major tactical maneuvers by which a rat can withdraw the nape and simultaneously turn to face the partner during play fighting. Either the defender rotates around its longitudinal axis to lie on its back, or it rotates but maintains firm hindpaw contact with the ground (Pellis et al. 1992). In both cases, the defender can then use its forepaws to block further attacks by the partner. However, the partial rotation tactic is more difficult to execute since the rat has to support its forequarters off the ground while making hindlimb adjustments to counter the maneuvers of the opponent, whilst simultaneously blocking with its forepaws. Rats using this maneuver are more likely to be knocked over by the attacker, and so are less likely to successfully defend the nape (Pellis & Pellis 1987; 1997b). As adults, it is the partial rotation tactic that is used most often in more agonistic encounters; yet as juveniles it is the complete rotation tactic that is the one most often used (Pellis & Pellis 1990). This example also belies another possible practice function of play fighting.

That is, it is used to enhance the ability to maintain the center of gravity during combat, rather than improve specific tactics. Again, the partial rotation tactic affords more opportunity for such training than the complete rotation tactic. Yet this is not the one most often used during juvenile play fighting.

If the practice hypothesis, as expressed by Symons (1978a) were correct, then it would be predicted that the most difficult tactics should receive more practice. This is clearly the expectation derived from the human motor learning literature (see above). As the examples discussed above show, this is not the case. The most difficult to execute tactics are not necessarily those receiving the most practice. However, it may be argued in the defense of the practice hypothesis that even poor practice is better than no practice. A closer examination of what differentiates play fighting from serious fighting further weakens this argument.

The role of play signals

Play fighting has been defined 'operationally in a given species by the presence of play signals and by the absence of agonistic communication' (Fagen 1981, p. 48). Presumably, play signals inform play partners that the interaction is playful (Bekoff 1975; Bekoff & Allen, Chapter 5.). As already noted, the supposed absence of threat signals during play fighting has been taken to be a design feature consistent with the practice hypothesis (Symons 1978a). Both of these operationally defined criteria of play fighting are, however, not universally applicable to either all species or to all play fights in a given species. There are many examples of threat signals being used during play fighting in a variety of species (see Bekoff 1995; Bekoff & Allen, Chapter 5; Pellis & Pellis 1996, and references therein). For example, in muntjacs, *Muntiacus reevesi*, playful sparring is initiated by antler pointing, which in other contexts is a serious threat (Barrette 1977). Therefore, even though threat signals may be absent from the play fighting of some species (e.g., rhesus monkeys, *Macaca mulatta,* Symons 1974), this may not generally be the case for other species. This weakens a supposed design feature of play fighting, one which has been argued in support of the practice hypothesis. However, it is possible that some species use play fighting as a means of practicing the use of threat signals. Symons (1978b) has provided a strong critique against this hypothesis, and it should be remembered that the presence of threat signals in social play has been thought to be rare (Fagen 1981). To our knowledge, no studies are available that demonstrate, either through experimentation or from an analysis of design features, the validity of such a functional hypothesis. Further analysis of this topic is beyond the scope of this chapter. It is suffice to say that the supposed absence of threat signals in play fighting has been regarded as a design feature supporting the hypothesis that, in play fighting, it is combat skills that are practiced (Symons 1974, 1978b). If threat signals are now found to be widespread in play fighting (Pellis & Pellis 1996), then support for the combat skill practice hypothesis is diminished. The presence of agonistic signals in play fighting also highlights our difficulty in identifying how interactants distinguish playful from agonistic encounters.

The ambiguity caused by the similarity of playful and serious fighting and by the presence of agonistic signals would seem to add to the value of play signals as markers of play. Unfortunately, play signals have not been identified in all species that play fight, and when present, appear to be used in only a minority of play fights (Pellis & Pellis 1997a). When available, play signals appear to be used in situations of ambiguity (Bekoff 1995). In most cases, the interactants appear to differentiate playful from serious fighting by contextual and stylistic cues (Pellis & Pellis 1996), with age, sex and individual identity of the play partner being some obvious contextual cues (Hayaki 1985). With regard to the design features of play fighting, we shall focus on the stylistic differences between playful and serious fighting. That is, differences in timing, strength, rhythm (Schwartzman 1979) and variations in form and sequence (Bekoff 1977). From this perspective, the difference between playful and serious fighting is not so much in the actual behavior patterns performed, but in how they are performed.

Stylistic features differentiating playful from serious fighting

During serious fighting, the attacking animal has to balance simultaneously two conflicting needs: those of attacking to gain the 'desired' advantage, and those of defending against a potential retaliatory strike from the opponent. Meanwhile, the defending animal has to use all possible means to block the successful execution of an attack (Pellis 1997). In contrast, our studies show that during play fighting, the attacker does not simultaneously defend against retaliation, nor does the defender vigorously block all attacks. Given that the targets of attack and defense greatly constrain the tactics used (Pellis 1989), the comparison between serious and playful fighting needs to be carefully made.

As in other muroid rodents (Pellis 1997), the main targets of attack during serious fighting in golden hamsters, *Mesocricetus auratus* are the lower flanks and dorsum (Pellis & Pellis 1988b). A defensive tactic often used is for the defender to stand upright on its hind legs and maintain its own head oriented towards the opponent's head. From this position, the defender can use the forepaws and the threat of a retaliatory bite to the face to block the attacker's maneuvers to gain access to the targeted area. The attacker may also stand upright and push the defender. If the defender becomes overbalanced, and so vulnerable, the attacker may then lunge and bite (Pellis & Pellis 1988a). A slightly modified version of this tactic in hamsters is illustrated in Pellis (1997). An upright attacker lunged downward at the lower right flank of an upright defender, simultaneously thrusting his right forepaw towards the defender's face; this blocked the defender's capacity to retaliate by lunging downward at the attacker's face (see Fig. 6.2). A comparable situation in the play fighting of another species illustrates the difference in the use of offensive tactics during play. Two spider monkeys, *Ateles geoffroyi* sat facing each other during a playful encounter. One lunged downward to bite the partner's lower arm. The other then lunged downward and gave a retaliatory bite to the attacker's head (Fig. 6.3). When launching its attack, the attacker made no attempt to block the partner's capacity to retaliate. Comparable examples occur in diverse species. For example, we have observed such a situation in an Australian parrot, the little corella, *Cacatua sanguinea*. While standing facing each other, one parrot raised its right foot and the other parrot, its left foot. After grabbing at each other's raised feet, one of the parrots lunged down to peck the other's foot, whereupon its partner lunged and pecked the side of the attacker's head. As in the spider monkey example, the attacking parrot made no attempt to block its partner's capacity for retaliation. The most extreme example we have personally witnessed is in Tonkean macaques, *Macaca tonkeana* where the attacker approached a reclining monkey. The attacker then sat on its partner's thigh, with its back facing the partner's head, and bit it on the hindleg. By attacking in this manner, the attacker left its back exposed for a counterattack. Indeed, upon being bitten, the defender lunged upward and bit the attacker on the back of the head!

A comparison of playful and serious attack within the same species further illustrates the difference in the use of defense during serious fighting versus play fighting. During serious combat, male Richardson's

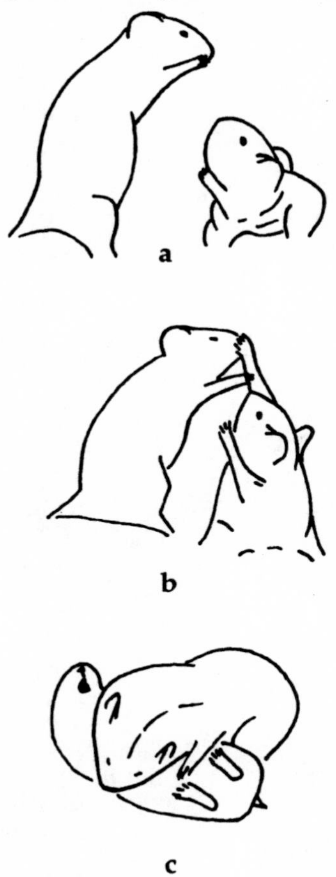

Fig 6.2. An example of combining defense with attack is shown for hamsters. From a lateral orientation, the attacking hamster (on the right) releases ground contact with its forepaws, and rotates its upper body so as to face the upright defender (a). The attacker then rises furthur upright, and uses its right forepaw to catch and hold the defender's head (b). The attacker then lungs and bites the flank, knocking the defender onto its back (c). By using the forepaw to block the defender's head, the attacker appears to prevent the defender from launching a retaliatory counterattack to its head, while simultaneously launching a bite to the defender's flank. (Modified from Pellis, 1997, © 1997, John Wiley & Sons, Inc).

ground squirrels *Spermophilus richardsonii* use a lateral attack maneuver in about 64% of encounters (Pellis et al. 1996). In contrast, during play fighting, juvenile males use this tactic in only 8% of encounters (MacDonald et al. 1994). In both forms of fighting, the attacker's attempt to gain contact of anterior body targets, and in both cases, retaliation can involve lunges at the face. When performing the lateral tactic during

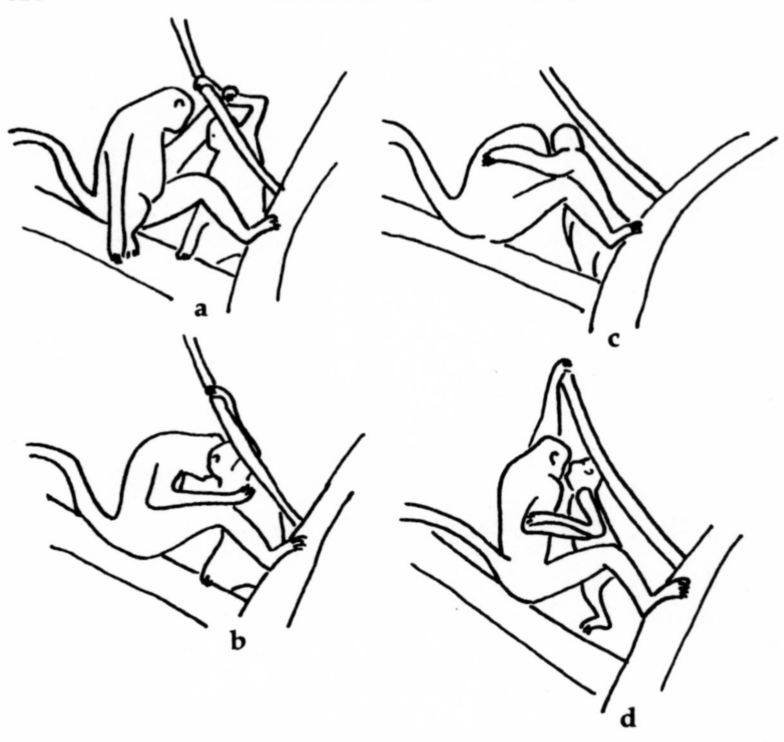

Fig 6.3. An example of the absence of combined defense with attack during play fighting is shown in two spider monkeys (on the left is an adult male, with a juvenile male on the right). From a face-to-face orientation (a), the adult male lunges down to bite the juvenile's right upper arm (b), the juvenile retaliates by lunging to bite the adult on the side of the face (c), which leads the adult to abort his attack and to defensively withdraw his head (d). Note that the adult's lunge was not combined with any restraining action to prevent a counterattack by the juvenile. (From Pellis and Pellis, 1997a © 1997, John Wiley & Sons, Inc).

serious fighting, the attacker stands in a broadside orientation towards the opponent and arches its back, usually piloerecting the tail. From this lateral position, the attacker can rapidly lunge at the opponent, or if the opponent retaliates, the attacker can pivot around its midbody and block the opponent's approach with a hip slam while at the same time withdrawing its head (Pellis et al. 1996). The lateral tactic thus serves as means of attack which simultaneously incorporates defense (Pellis 1997). Not only is the lateral tactic rare during play fighting, but when used, it is typically performed by the defending, not the attacking, squirrel. That is, in play fighting, attack and defense are rarely simultaneously executed by the attacker.

The vigor of defense also differs for playful and serious fighting. For example, given that in rats the targets of attack and defense differ between playful and serious fighting, the patterns of attack used in play fighting are unlike those used in serious fighting. However, even though modified, many of the tactics of defense are similar between the two forms of fighting. One such tactic involves turning around, while standing, to face the opponent, and counterattacking once the opponent contacts the rump area (see Fig. 6.1). In play fighting, the rump contact is transitory as the attacker shifts towards the nape (Pellis & Pellis 1987); whereas in serious fighting, the contact on the rump would result in a bite (Blanchard et al. 1977). The duration of this tactic, from the frame of first contact until the defending rat had turned its longitudinal body axis 180°, was calculated in seconds, and found to be significantly shorter in serious fighting than in play fighting ($\overline{X} \pm SE$: 0.14 ± 0.01 versus 0.21 ± 0.02; $t = 4.85$, $df = 34$, $p < 0.001$). This example graphically illustrates that playful defense is not as vigorous as serious defense, in that more opportunity is left for the successful prosecution of the partner's attack.

Though it is relatively rare, play fighting can sometimes escalate to serious fighting, especially with the onset of puberty (Fagen 1981). Following sexual maturity, escalation from playful to serious aggression can be as high as 30% in golden hamsters (Pellis & Pellis 1988a), to as low as 10% in rats (Pellis & Pellis 1987) and 1% in river otters, *Lutra canadensis* (Beckel 1991). In the juvenile phase, such escalation is even rarer, dropping to less than 1% in both rats and hamsters. Agonism escalating from play fighting was reported in 12% of encounters by juvenile male punarés, *Thrichomys apereoides*, but only between unrelated animals. Play fighting amongst juvenile male siblings was never observed to escalate to agonism (Thompson & Cranford 1985). Even though relatively rare, such escalation can be useful for analysing the stylistic differences between playful and serious fighting. By analysing the behavior immediately preceding such escalation, it was found that for both rats (Pellis & Pellis 1991) and hamsters (Pellis & Pellis 1988a), the intensity of the attack or the defense increased. In attack, the defender was typically held down by the attacker's forepaws – this strongly blocked the defender's ability to launch counterattacks. In defense, the attacker was held at bay by the defender's forepaws – this blocked the attacker's ability to contact the play target. In both cases, the restrained pairmate switched to launching serious attacks (i.e., bites to the rump). Therefore, escalation appears to occur when the attacker combines its attack with effective defense, or when the defender vigorously blocks all attacks. When this

Table 6.1. *Combat rules distinguishing playful from serious fighting.*

	Type of fighting	
Combat role	serious	playful
Attacker	simultaneous attack and defence (prevent counterattack)	sequential attack and defense (allow counterattack)
Defender	vigorous defense (prevent all attacks)	ineffective defense (allow some attacks)

happens, the interaction ceases to be playful, and the opponent switches to being aggressive. Similarly, Altman (1962) notes that unless both interactants have an equal opportunity to 'win', playful interactions by rhesus macaques are rapidly terminated. Aldis (1975) termed this need for both interactants to 'win' some play fights as the 50–50 rule. What distinguishes playful from serious fighting, then, is the style of combat (Table 6.1). In this scheme, play signals, when available, may be viewed as substituting or supplementing these stylistic features if some ambiguity arises (Pellis & Pellis 1996). Therefore, the categorization of play fighting as involving the rule – 'bite, but avoid being bitten' (Biben 1986) – is not an accurate reflection of the content of play fighting. Instead, the rule is 'bite, allow yourself to be bitten, then defend yourself, and bite again'!

This framework for differentiating play fighting from serious fighting is similar to the notions of self-handicapping and role reversal which are regarded as central to the playful mode of combat (Fagen 1981). During serious fighting, when both combatants are seeking to gain a victory, role reversals can also occur (e.g., Geist 1971; Pellis et al. 1996). What distinguishes the role reversals in play fighting is that the interactants actually provide the opportunity for the partner to gain an advantage. In serious fighting, such a reversal only occurs by successfully overpowering the opponent's defenses (Pellis 1997). Typically, self-handicapping has only been reported in the context of a dominant, older, or larger individual modifying its behavior so as to give the competitive advantage to its partner (e.g., Biben 1989; LeResche 1976; Symons 1978a; Watson & Croft 1996). In the framework presented above, self-handicapping is a pervasive feature of all play fighting, which may become more exaggerated with more extreme asymmetries between partners.

With regard to the practice hypothesis, it is argued here that the organizational style of play fighting itself is antithetical to the practice of combat tactics. That is, for play fighting to remain playful, it cannot practice the behavior patterns most useful in serious combat. The most difficult maneuvers used by an attacker during serious fighting are those requiring combined attack and defense and the most difficult task for a defender is to counter every move by the attacker. These are not the patterns of action used during play fighting. If, as noted above, the human motor learning literature is taken as a standard for the evaluation of practice, then the actual sequences of movement that need to be enhanced need to be performed (Stamps 1995). In this regard, play fighting fails to provide the practice needed.

A possible criticism of this view is that juveniles do not yet possess the full adult repertoire, and hence cannot perform the most adult-like patterns of attack and defense. Therefore, using what they have is the best they can do in terms of practice, and again, this may be better than nothing. The empirical developmental data available are not consistent with this criticism. Young coyotes, *Canis latrans* fight seriously to establish dominance relationships prior to the onset of juvenile play fighting (Bekoff 1978). Even more dramatically, spotted hyenas, *Crocuta crocuta* engage in highly escalated combat immediately following birth. In many cases, this leads to the death of one of the combatants, and only later in development is serious fighting replaced by play fighting (Drea et al. 1996). In rats, playful defense in infancy involves the same tactics as are used post-pubertally, but during the juvenile phase, they switch to the juvenile-typical pattern of rotating to supine (Pellis & Pellis 1997b). In all these cases, the juveniles already have the capability of engaging in the tactics of serious fighting. That during the juvenile phase they do not use these tactics in the combat-typical style during play fighting is evidence that the form of play fighting is different to that of serious fighting, and is not simply a byproduct of incomplete maturation. For the argument developed in this paper, this means that if play fighting were present for the rehearsal of combat tactics, then juveniles should use the actual tactics of combat, as these are present in their repertoire. The fact that they do not do so, or do so only to a limited degree, supports the view that practicing combat tactics is unlikely to be the main function that has shaped the structure of play fighting.

Combat relevant and combat irrelevant behavior patterns

Whether playful or serious, fighting involves the performance of maneuvers that are used to gain an advantage by one opponent and those that are used by the defender to deny an opponent that advantage. In serious fighting, successful use of attack and defense is crucial; if an opponent's guard is let down, it could suffer serious or even lethal injury (Geist 1971). As seen above, in play fighting, such defense is not crucial, and indeed, is counter to the playful character of the interaction. Nonetheless, most of the content of play fighting involves tactics of attack and defense. Movements and postures that are irrelevant to attack and defense during serious fighting are the gestures of threat, dominance or submission (Pellis 1997). These behavior patterns are typically derived from the tactics of fighting, and function to achieve the consequences of fighting (Walther 1974). While such gestures may sometimes occur in play fighting (Pellis & Pellis 1996), other kinds of combat unrelated movements also occur.

Some of these irrelevant movements, when they occur in young animals, may be due to lack of maturation of the central control mechanisms over motor output (Hogan 1988). For example, just after weaning, when play fighting emerges as a dominant pattern of interaction amongst young rats (Bolles & Woods 1964), the pups perform jerky jumping movements that may interfere with effective attack and defense (Pellis & Pellis 1983). More intriguingly, there are cases when two animals are engaged in play fighting and one switches to some other activity, even though the partner is still attacking. For example, two Australian magpies, *Gymnorhina tibicen* lying side-by-side on the ground while facing each other, grapple with their feet and peck with their beak. While still grappling with each other's feet, one magpie reaches over to a stick, picks it up in the beak and mandibulates it playfully (Pellis 1981b). Clearly, this stick oriented behavior is irrelevant to the play fight, and indeed, leaves its play target, the side of the head, exposed to attack by the partner.

As with the rat example, the young magpies may be exhibiting properties of incomplete maturation, where they can be easily distracted from the immediate task by some novel stimulus. Such a possibility is illustrated by the development of foraging behavior in these birds. By about 8 weeks after fledging, young magpies appear similar to the adults in their foraging behavior. They walk slowly over an open field, striking at and catching small arthropod prey at the same rate as the adults; the capture success rate is 78% for adults and 75% for juveniles. Furthermore, the size and quality of the prey are similar. On one afternoon, a swarm of beetles filled the air. The adults did not change their foraging strategy. They continued to walk slowly and to catch the beetles that fell to the ground. In contrast, the young magpies ran after one beetle, but as soon as it dropped in the grass, another that flew passed was chased, and so on. The juveniles were clearly distracted by, and attracted to, movement. While the adult success rate in capturing prey remained at 78%, that of the young dropped to about 54% (Pellis 1981a). Therefore, it is possible that some of the irrelevant actions during play fighting are byproducts of other neurobehavioral characteristics of infants and juveniles, rather than being intrinsic properties of play. Indeed, with age, the jerky jumps of rats become less inappropriate (Pellis & Pellis 1983), and are then only used as aids for attack and defense (Pellis & Pellis 1987).

We have personally witnessed only one species in which an animal seems to 'deliberately' perform an action that impedes its ability to attack during play fighting. Observations of a captive troop of Western lowland gorillas, *Gorilla gorilla gorilla* containing two 2 year old infants showed that they conformed to the basic pattern of play fighting previously described for gorillas. That is, in gorillas, play fighting involves face-to-face grappling and slapping, with bites directed at the side of the neck (Schaller 1963). However, on two occasions, one infant approached the other, who was standing behind a fire hose that was strung low between two trees. The attacking infant gripped the hose in his teeth and then grappled and slapped at the other infant. The defender grappled and slapped in return, and made biting lunges at the side of the other's neck. Meanwhile, the attacker warded off these counterattacks with his hands while still holding onto the hose (unpublished observations). Clearly, the attacker engaged in movements that were not relevant to combat, and indeed, were even counterproductive to gaining an advantage over the partner (see also Aldis 1975, for observations on the common chimpanzee). While this example is reminiscent of the self-handicapping discussed above, there are important differences. In self-handicapping, the performer typically handicaps itself in a way so as to entice an attack by its partner; for example, an adult male Hamadryas baboon, *Papio hamadryas* may roll onto its back in front of a juvenile (LeResche 1976). However, once on its back, the initiator is still in a position to defend itself and launch counterattacks if the partner attacks. The gorilla case, on the other hand, is more bizarre in that the attacker handicaps himself by preventing an effective attack *while* attacking.

The presence of irrelevant movements, whether involuntary or deliberate, makes the play fighting of young animals an inappropriate vehicle for practicing fighting skills. Again, the detailed content of play fighting is inconsistent with the practice hypothesis.

Can play fighting be used for practice?

It should not be concluded, however, that play fighting can never be used for practicing motor skills. Sometimes the correlation between performance and age of maximal benefit are so close (Fagen 1980) that the practice hypothesis would seem very plausible. For example, Australian magpies use coordinated foot and beak movements to fragment large prey. The prey, held in the beak, is grasped by a foot, and held on the ground; then, by upward and lateral head movements, pieces are pulled off. The most difficult part of the sequence involves grasping the object held in the beak with the foot. Detailed movement notation analysis of the orchestration of the body and limb movements revealed that the foot is flicked upward and forward to about the same location in space irrespective of the size of the object, and that successful contact required coordinated head and body movements which brings the beak downward to that location (Pellis 1983). During juvenile play with objects, these same maneuvers are used. However, in the first three weeks following fledging, all such leg flicks are unsuccessful as the head and body movements fail to be appropriately coordinated to the leg movements. By approximately four weeks after fledging, as the head, body and leg movements become coordinated, about 52% of these leg flicks are successful (Fig. 6.4). Then, over the next four weeks, the success rate gradually increases to over 90%. Corresponding to these changes in motor coordination is a marked change in the frequency of use. From the first three weeks following fledging, leg flicks occur at a rate of about one per minute of object manipulation, then at four weeks, when the success rate abruptly increases, the rate jumps to over 5 per minute. Then, as the rate of success gradually increases in the following weeks, the rate declines to about 3.5 per minute (Pellis 1983). That is, the repetition rate is highest when the change in coordination is the greatest, exactly what would be predicted by the practice hypothesis.

With regard to play fighting, we suggest that if practice does occur, it is most likely to be for defense, not attack. By incorporating defense sequentially rather than simultaneously with playful attack, the structure of playful attack only superficially resembles that of serious attack. In contrast, the tactics of defense are more closely matched motorically in the two forms of fighting, with differences being primarily in timing and rhythm. An example of possible motor practice via play fighting is illustrated in rats. When an adult rat is eating a food item held in its paws, another rat may approach from the side to rob the item. The eating rat defends the food item by swerving laterally away from the attacker (Whishaw 1988). While both males and females perform this defensive dodging, they use a different combination of movements (Field et al. 1996). While females pivot around their pelvis, males pivot around the midbody. Thus for females, the whole body is moved unidirectionally away from the robber, whereas for males, the forequarters move away, with the hindquarters moving towards the robber. The female tactic is fully formed from its first appearance shortly after weaning, whereas the male tactic lacks coordination of the movements of the fore- and hindquarters. This results in a staccato sequence of movement as the fore-

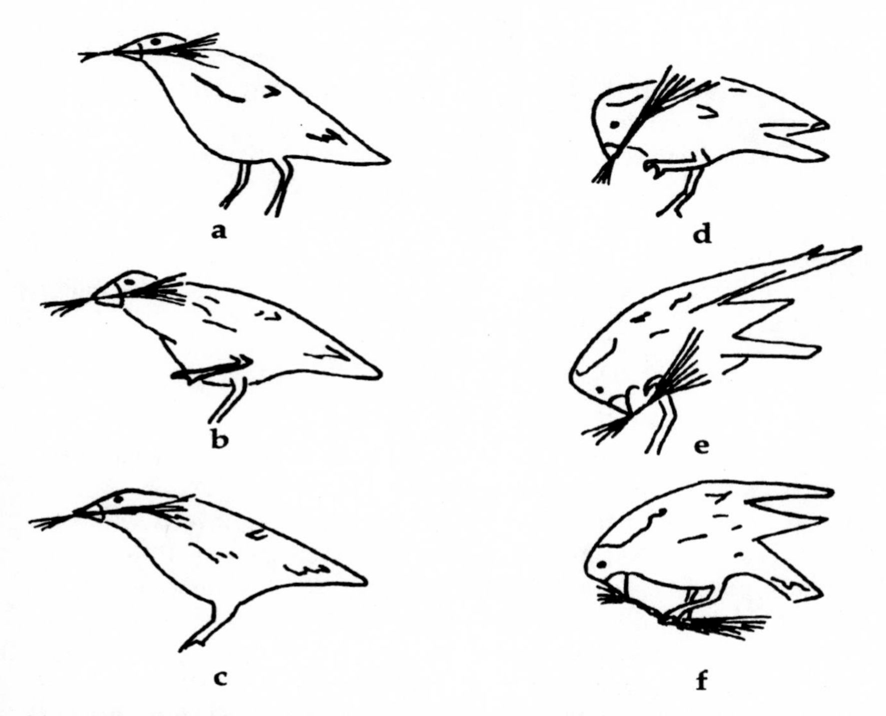

Fig 6.4. An example of an unsuccessful and a successful grasp by the foot with an object (a pine twig) held in the bill by a juvenile Australian magpie. In the unsuccessful attampt, the magpie fails to lower its head and neck sufficiently for the foot to reach the twig (a-c). In the successful reach, the magpie's head is rotated downwards (d), and it's body is tilted downward anteriorly, which enables it to grasp (e) and hold (f) the twig with the foot. The kinematic structure of the leg movements are the same in both attempts. However, in the second case, the movements of the head and body are coordinated with those of the leg, allowing successful contact. (Modified from Pellis 1983 © 1983, John Wiley & Sons, Inc).

quarters are moved a little, then the hindquarters are moved, and so on. The male pattern is fully formed shortly after sexual maturity (Field 1996). In play fighting, a similar lateral dodge is performed by rats in order to defend the nape from a lateral attack (Pellis & Pellis 1990), with similar sex differences in motor organization (Pellis et al. 1997).

In an isolation experiment, from weaning until adulthood, pairs of male rats were housed in cages with a wire mesh partition between them. The pairs of rats were then housed together for four weeks prior to testing in the food robbing and dodging paradigm. The socially isolated rats, now 150 days old, still had the staccato dodging movements typical of late infancy (Field et al., 1997). Indeed, even after repeated trials, they did not seem to improve their fore- and hindquarter coordination. Given that as juveniles these rats failed to engage this response either to defend a food item or their napes during play fighting, the lack of coordination cannot be causally connected to either source of experience alone. Nonetheless, even though dodging comprises only about 20% of defensive maneuvers during play fighting (Pellis & Pellis 1990), during the whole of the juvenile phase, this maneuver is performed hundreds of times. Therefore, this example at least suggests that juvenile experience with motor performance, especially in play, may influence later motor execution, at least for some defensive tactics.

Conclusion

Both the behavior patterns used and the style of their use makes play fighting a poor means by which to practice fighting skills. At a superficial level, playful fighting resembles serious fighting. Both involve competition for attaining some advantage (e.g., Aldis 1975; Symons 1978a). This superficial resemblance probably led to the practice hypothesis in the first place. Many of the design features of play fighting that have been argued to support this hypothesis are crude correlations. Sex differences (Fagen 1981; Smith 1982), terrain preferences (Byers 1980) and the absence of threat signals (Symons 1978a), are either of limited validity (Pellis & Pellis 1996), or are unequivocal design features for this proposed function (Coppinger & Smith 1989; Pellis et al. 1997). Furthermore, some other features of play fighting, such as the occurrence of combat unrelated movements and altered sequencing, are counter to its supposed function of practicing combat skills (Loizos 1966).

For a functional hypothesis to be convincing, it must predict the design features of the trait (Williams 1966). With respect to its detailed structure, play fighting does not appear to be useful for practice, in that it does not refine combat tactics. Practice means rehearsing specific tactics, and then incorporating them into larger sequences. Performing actions that may be superficially similar is not practice. In particular, the design requirements of playful attack (Table 6.1) make it poorly correlated to the motor demands of serious attack. As noted above, there is a greater likelihood that if such practice occurs, it would be for defense.

Therefore, while it is possible that some aspects of combat may be rehearsed during play fighting, the present analysis suggests that the practice hypothesis cannot account for the majority of the design features of play fighting. Other variations of this practice or training hypothesis require similar detailed analyses of the structure of play fighting for their evaluation (Byers & Walker 1995). The design features of play fighting outlined in this paper suggest two alternative functions for play fighting.

A major problem in initiating a serious attack on an opponent is to overcome the fear of retaliation (Pellis 1997). Even though the attacker may take precautions (i.e., simultaneous defense with offense), the attacker still risks being injured by a successful retaliation (Pellis & Pellis 1992). In play fighting, sequential attack and defense guarantees that the partner has the opportunity to counterattack successfully. Therefore, one possibility is that play fighting functions as a means of reducing the fear of launching attacks. A similar suggestion has been made for predatory play (Martin & Caro 1985), where playful exposure to a live prey may diminish the fear of such prey. Indeed, naive adult cats, *Felis catus* that are poor killers can be induced into more effective predatory attack by injections of benzodiazepines (Pellis et al. 1988; see also Siviy, Chapter 11, for a general discussion of neurobiological analyses of play). The drug improves predatory attack by reducing the cat's fear of the prey, rather than by improving its predatory skills. More generally, given that during play fighting animals are rough-housed, and sometimes even hurt, such an activity may be useful in training the animals to accept some degree of pain during social interactions (Potegal & Einon 1989).

Another possible function of play fighting is that it serves some immediate social goals. Repeated engagement in this pseudo-agonistic pattern of interaction may provide the participants with detailed knowledge of any alterations in their relationship (Wolf 1984), reaffirm their relationship (Pellis et al. 1993), or test it (Croft & Snaith 1991; Paquette 1994). For example, in children, the intensity and roughness of play fighting appears to be modulated depending upon whether it is being used to reaffirm friendships or to jockey for position in a dominance hierarchy (Neill 1976; Pellegrini 1988; Smith & Boulton 1990).

Whether these alternative hypotheses for the functions of play fighting are correct will depend upon a demonstrated concordance between the fine structure of play fighting and the purported function(s). That is, description is suggestive of a different set of hypotheses for the function of play fighting, and these hypotheses will generate finer grained descriptions that in turn will test the validity of the hypotheses.

Acknowledgements

We thank M. Bekoff, M. Biben, J. Byers and R. Fagen for their valuable comments and Adria Allen for typing the paper. Much of the work reported herein was supported by grants from the Harry Frank Guggenheim Foundation and the Natural Sciences and Engineering Research Council of Canada to S. M. P.

References

Aldis, O. 1975. *Play Fighting*. New York: Academic Press.

Altmann, S. A. 1962. Social behavior of anthropoid primates: Analysis of recent concepts. In: *Roots of Behavior* (ed. E. L. Bliss), pp. 277–85. New York: Harper & Brothers Publishers.

Alvarez, F. 1993. Risks of fighting in relation to age and territory holding in fallow deer. *Can. J. Zool.*, **71**, 376–83.

Angier, N. 1995. *The Beauty of the Beastly*. Boston: Houghton Mifflin.

Barrette, C. 1977. The social behaviour of captive muntjacs *Muntiacus reevesi* (Ogilby 1839). *Z. Tierpsychol.*, **43**, 188–213.

Beckel, A. L. 1991. Wrestling play in adult river otters, *Lutra canadensis*. *J. Mammal.*, **72**, 386–90.

Bekoff, M. 1975. The communication of play intention: Are play signals functional? *Semiotica*, **15**, 231–39.

Bekoff, M. 1977. Social communication in canids: Evidence for the evolution of a stereotyped mammalian display. *Science*, **197**, 1097–9.

Bekoff, M. 1978. Behavioral development in coyotes and Eastern coyotes. In: *Coyotes: Biology, Behavior and Management* (ed. M. Bekoff), pp. 97–126. New York: Academic Press.

Bekoff, M. 1995. Play signals as punctuation: The structure of social play in canids. *Behaviour*, **132**, 419–29.

Berger, J. 1980. The ecology, structure and function of social play in bighorn sheep; (*Ovis canadensis*). *J. Zool.*, **192**, 531–42.

Biben, M. 1982. Sex differences in the play of young ferrets. *Biol. Behav.* **7**, 303–8.

Biben, M. 1986. Individual- and sex-related strategies of wrestling play in captive squirrel monkeys. *Ethology*, **71**, 229–41.

Biben, M. 1989. Effects of social environment on play in squirrel monkeys: resolving Harlequin's dilemma. *Ethology*, **81**, 72–82.
Bolles, R. C. & Woods, P. J. 1964. The ontogeny of behavior in the albino rat. *Anim. Behav.*, **12**, 427–41.
Byers, J. 1980. Play partner preferences in Siberian ibex, *Capra ibex sibirica. Z. Tierpsychol.*, **53**, 23–40.
Byers, J. A. & Walker, C. 1995. Refining the motor training hypothesis for the evolution of play. *Am. Nat.*, **146**, 25–40.
Caro, T. M. 1988. Adaptive significance of play: Are we getting closer? *Trends Ecol. Evol.*, **3**, 50–4.
Coppinger, R. P. & Smith, C. K. 1989. A model for understanding the evolution of mammalian behavior. In: *Current Mammalogy (Vol. II)* (ed. H. Genoways), pp. 53–73. New York: Plenum Press.
Croft, D. B. & Snaith, F. 1971. Boxing in red kangaroos, *Macropus rufus*: Aggression or play? *Int. J. Comp. Psych.*, **4**, 221–36.
Crowell-Davis, S. L., Houpt, K. A. & Kane, L. 1987. Play development in Welsh pony (*Equus caballus*) foals. *Appl. Anim. Behav. Sci.*, **18**, 119–31.
Drea, C. M., Hawk, J. E. & Glickman, S. E. 1996. Aggression decreases as play emerges in infant spotted hyaenas: Preparation for joining the clan. *Anim. Behav.*, **51**, 1323–36.
Einon, D., Morgan, M. & Kibbler, C. C. 1978. Brief periods of socialization and later behaviour in the rat. *Dev. Psychobiol.*, **11**, 213–25.
Fagen, R. 1980. Ontogeny of animal play behavior: Bimodal age schedules. *Anim. Behav.*, **28**, 1290.
Fagen, R. 1981. *Animal Play Behaviour*. Oxford, UK: Oxford University Press.
Field, E. F. 1996. *Sex Differences in Movement Organization*. Unpublished Master's dissertation, University of Lethbridge.
Field, E. F., Whishaw, I. Q. & Pellis, S. M. 1996. A kinematic analysis of evasive dodging movements used during food protection in the rat: Evidence for sex differences in movement. *J. Comp. Psychol.*, **110**, 298–306.
Field. E. F., Whishaw, I. Q., & Pellis, S. M. 1997. Lack of experience during the juvenile phase does not affect the development of sex-typical movement patterns. Presented at the Society for Neuroscience 27th annual meeting. New Orleans. November.
Fox, M. 1969. The anatomy of aggression and its ritualization in Canidae: A developmental and comparative study. *Behaviour*, **35**, 242–58.
Geist, V. 1971. *Mountain Sheep*. Chicago: The University of Chicago Press.
Groos, K. 1898. *The Play of Animals*. New York: Appleton.
Hård, E. & Larsson, K. 1971. Climbing behavior patterns in prepubertal rats. *Brain Behav. Evol.*, **4**, 151–61.
Hayaki, H. 1985. Social play of juvenile and adolescent chimpanzees in the Mahale Mountains National Park, Tanzania. *Primates*, **26**, 343–60.
Hogan, J. A. 1988. Cause and function in the development of behavior systems. In: *Handbook of Behavioral Neurobiology: Vol. 9* (ed. E. M. Blass), pp. 63–106. New York: Plenum Press.
Hole, G. & Einon, D. F. 1984. Play in Rodents. In: *Play in Animals and Man* (ed. P. K. Smith), pp. 95–117. Oxford, UK: Basil Blackwell.
Larsson, K. 1978. Experiential factors in the development of sexual behaviour. In: *Biological Determinants of Sexual Behaviour* (ed. J. B. Hutchinson), pp. 55–86. New York: John Wiley and Sons.

LeResche, L. A. 1976. Dyadic play in hamadryas baboons. *Behaviour,* **57**, 190–205.

Loizos, C. 1966. Play in mammals. *Symp. Zool. Soc. Lond.,* **18**, 1–9.

MacDonald, N. L., Pellis, S. M. & Michener, G. R. 1994. Sex differences in the play behavior of juvenile Richardson's ground squirrels (*Spermophilus richardsonii*). Poster presented at the 31st Annual Meeting of the Animal Behaviour Society, Seattle, WA, July.

Martin, P. & Caro, T. M. 1985. On the functions of play and its role in behavioral development. *Adv. Study Behav.,* **15**, 59–103.

McClintock, M. & Adler, N. T. 1978. The role of the female during copulation in wild and domestic rats (*Rattus norvegicus*). *Behaviour,* **67**, 67–96.

Meaney, M. J., Stewart, J. & Beatty, W. W. 1985. Sex differences in social play: The socialization of sex roles. *Adv. Study Behav.,* **15**, 1–58.

Moore, C. L. 1985. Development of mammalian sexual behavior. In: *The Comparative Development of Adaptive Skills: Evolutionary Implications* (ed. E. S. Gollin), pp. 19–56. Hillsdale, NJ: Lawrence Erlbaum Assoc., Inc.

Moran, G. 1985. Behavioral description and its impact on functional inference. *Behav. Brain Sci.,* **8**, 186–7.

Neill, S. R. 1976. Aggressive and non-aggressive fighting in 12–13 year old preadolescent boys. *J. Child Psychol. Psychiat.*, **17**, 213–20.

Panksepp, J. 1981. The ontogeny of play in rats. *Dev. Psychobiol.*, **14**, 327–32.

Paquette, D. 1994. Fighting and play fighting in captive adolescent chimpanzees. *Aggress. Behav.*, **20**, 49–65.

Pellegrini, A. D. 1988. Elementary school children's rough-and-tumble play and social competence. *Dev. Psychol.*, **24**, 802–6.

Pellis, S. M. 1981a. Exploration and play in the behavioral development of the Australian magpie *Gymnorhina tibicen. Bird Behav.,* **3**, 37–49.

Pellis, S. M. 1981b. A description of social play by the Australian magpie *Gymnorhina tibicen* based on Eshkol-Wachman movement notation. *Bird Behav.,* **3**, 61–79.

Pellis, S. M. 1983. Development of head and foot coordination in the Australian magpie *Gymnorhina tibicen* and the function of play. *Bird Behav.,* **4**, 57–82.

Pellis, S. M. 1985. What is 'fixed' in a fixed action pattern? A problem of methodology. *Bird Behav.,* **6**, 10–15.

Pellis, S. M. 1988. Agonistic versus amicable targets of attack and defense: Consequences for the origin, function and descriptive classification of play-fighting. *Aggress. Behav.*, **14**, 85–104.

Pellis, S. M. 1989. Fighting: the problem of selecting appropriate behavior patterns. In: *Ethoexperimental Approaches to the Study of Behavior* (eds. R. J. Blanchard, P. F. Brain, D. C. Blanchard & S. Parmigiani), pp. 361–74. Dordrecht, The Netherlands: Kluwer Academic Publishers.

Pellis, S. M. 1993. Sex and the evolution of play fighting: A review and model based on the behavior of muroid rodents. *Play Theory Res.*, **1**, 55–75.

Pellis, S. M. 1997. Targets and tactics: The analysis of moment-to-moment decision making in animal combat. *Aggress. Behav.*, **23**, 107–29.

Pellis, S. M. & Pellis, V. C. 1983. Locomotor-rotational movements in the ontogeny and play of the laboratory rat *Rattus norvegicus. Dev. Psychobiol.*, **16**, 269–86.

Pellis, S. M. & Pellis, V. C. 1987. Play-fighting differs from serious fighting in both target of attack and tactics of fighting in the laboratory rat *Rattus norvegicus. Aggress. Behav.,* **13**, 227–42.

Pellis, S. M. & Pellis, V. C. 1988a. Play fighting in the Syrian golden hamster (*Mesocricetus auratus* Waterhouse) and its relationship to serious fighting during post-weaning development. *Dev. Psychobiol.*, **21**, 323–37.
Pellis, S. M. & Pellis, V. C. 1988b. Identification of the possible origin of the body target which differentiates play-fighting from serious fighting in Syrian golden hamsters *Mesocricetus auratus. Aggress. Behav.*, **4**, 437–49.
Pellis, S. M. & Pellis, V. C. 1989. Targets of attack and defense in the play fighting by the Djungarian hamsters *Phodopus campbelli*: Links to fighting and sex. *Aggress. Behav.*, **15**, 217–34.
Pellis, S. M. & Pellis, V. C. 1990. Differential rates of attack, defense and counterattack during the developmental decrease in play fighting by male and female rats. *Dev. Psychobiol.*, **23**, 215–31.
Pellis, S. M. & Pellis, V. C. 1991. Role reversal changes during the ontogeny of play fighting in male rats: Attack versus defense. *Aggress. Behav.*, **17**, 179–89.
Pellis, S. M. & Pellis, V. C. 1992. An analysis of the targets and tactics of conspecific and predatory attack in Northern grasshopper mice *Onychomys leucogaster. Aggress. Behav.*, **18**, 301–16.
Pellis, S. M. & Pellis, V. C. 1996. On knowing it's only play: The role of play signals in play fighting. *Aggress. Violent Behav.*, **1**, 249–68.
Pellis, S. M. & Pellis, V. C. 1997a. Targets, tactics and the open mouth face during play fighting in three species of primates. *Aggress. Behav.*, **23**, 41–57.
Pellis, S. M. & Pellis, V. C. 1997b. The pre-juvenile onset of play fighting in laboratory rats *Rattus norvegicus. Dev. Psychobiol.* **31**, 193–205.
Pellis. S. M., Whishaw. I. Q, & Pellis. V. C. 1992. The role of the cortex in the play fighting by rats. Developmental and evolutionary implications. *Brain. Behav. Evol.*, **39**, 270–84.
Pellis, S. M., MacDonald, N. L. & Michener, G. R. 1996. Lateral display as a combat tactic in Richardson's ground squirrel *Spermophilus richardsonii. Aggress. Behav.*, **22**, 119–34.
Pellis, S. M., Pellis, V. C. & McKenna, M. M. 1993. Some subordinates are more equal than others: Play fighting amongst adult subordinate male rats. *Aggress. Behav.*, **19**, 385–93.
Pellis, S. M., Pellis, V. C. & McKenna, M. M. 1994. A feminine dimension in the play fighting of rats (*Rattus norvegicus*) and its defeminization neonatally by androgens. *J. Comp. Psychol.*, **108**, 68–73.
Pellis, S. M., Field, E. F., Smith, L. K. & Pellis, V. C. 1997. Multiple differences in the play fighting of male and female rats. Implications for the causes and functions of play. *Neurosci. Biobehav. Rev.*, **21**, 105–20.
Pellis, S. M., O'Brien, D. P., Pellis, V. C., Teitelbaum, P., Wolgin, D. L. & Kennedy, S. 1988. Escalation of feline predation along a gradient from avoidance through 'play' to killing. *Behav. Neurosci.*, **102**, 760–77.
Pfeiffer, S. 1985. Sex differences in social play of scimitar-horned oryx calves (*Oryx dammah*). *Z. Tierpsychol.*, **69**, 281–82.
Poole, T. B. 1966. Aggressive play in polecats. *Symp. Zool. Soc. Lond.*, **18**, 23–44.
Poole, T. B. & Fish, J. 1975. An investigation of playful behaviour in *Rattus norvegicus and Mus musculus* (Mammalia). *J. Zool.*, **175**, 61–71.
Potegal, M. & Einon, D. 1989. Aggressive behaviors in adult rats deprived of playfighting experience as juveniles. *Dev. Psychobiol.*, **22**, 159–72.

Rhine, R. J. 1973. Variation and consistency in the social behavior of two groups of stumptail macaques (*Macaca arctoides*). *Primates*, **14**, 21–35.

Schaller, G. B. 1963. *The Mountain Gorilla*. Chicago: The University of Chicago Press.

Schwartzman, H. B. 1979. *Transformations: The Anthropology of Children's Play*. New York: Plenum Press.

Smith, P. K. 1982. Does play matter? Functional and evolutionary aspects of animal and human play. *Behav. Brain Sci.*, **5**, 139–84.

Smith, P. K. 1985. Functional hypothesis and their impact on behavioral description. *Behav. Brain Sci.*, **8**, 187–8.

Smith, P. K. & Boulton, M. J. 1990. Rough-and-tumble play, aggression and dominance: perception and behaviour in children's encounters. *Human Develop.*, **33**, 271–82.

Stamps, J. 1995. Motor learning and the value of familiar space. *Am. Nat.*, **146**, 41–58.

Symons, D. 1974. Aggressive play and communication in rhesus monkeys (*Macaca mulatta) Am. Zool.*, **14**, 317–22.

Symons, D. 1978a. *Play and Aggression: A Study of Rhesus Monkeys*. New York: Columbia University Press.

Symons, D. 1978b. The question of function: Dominance and play. In: *Social Play in Primates* (ed. E. O. Smith), pp. 193–230. New York: Academic Press.

Takahashi, L. K. & Lore, R. K. 1983. Play fighting in juvenile rats and the development of agonistic behavior in male and female rats. *Aggress. Behav.*, **9**, 217–27.

Taylor, G. T. 1980. Fighting in juvenile rats and the ontogeny of agonistic behavior. *J. Comp. Physiol. Psychol.*, **94**, 953–61.

Thompson, K. V. & Cranford, J. A. 1985. Social play and factors precipitating escalation into agonism in juvenile punarés (*Thrichomys apereoides*). *Am. Zool.*, **25**, 77A.

Walther, F. 1974. Some reflections on expressive behavior in combat and courtship of certain horned ungulates. In: *The Behavior of Ungulates and Its Relation to Management* (eds. V. Geist & F. Walther), pp. 56–106. Morges, Switzerland: IUCN Publication No. 24.

Watson, D. M. & Croft, D. G. 1993. Play fighting in captive red-necked wallabies *Macropus rufogriseus banksianus. Behaviour*, **125**, 219–45.

Watson, D. M. & Croft, D. B. 1996. Age-related differences in play fighting strategies of captive male red-necked wallabies (*Macropus rufogriseus banksianus*). *Ethology*, **102**, 336–46.

Whishaw, I. Q. 1988. Food wrenching and dodging: Use of action patterns for the analysis of sensorimotor and social behavior in the rat. *J. Neurosci. Methods*, **24**, 169–78.

Whishaw, I. Q. & Whishaw, G. 1996. Conspecific aggression influences food carrying: Studies on a wild population of *Rattus norvegicus*. *Aggress. Behav.*, **22**, 47–66.

Williams, G. C. 1966. *Adaptation and Natural Selection*. Princeton, NJ: Princeton University Press.

Wolf, D. P. 1984. Repertoire, style and format: Notions worth borrowing from children's play. In: *Play in Animals and Humans* (ed. P. K. Smith), pp. 175–93. Oxford, U.K.: Basil Blackwell.

7

Sparring as play in young pronghorn males

MICHELLE N. MILLER
Biology Department, Blue Mountain Community College, Pendleton, OR 97801 USA

and JOHN A. BYERS
Department of Biological Sciences, University of Idaho, Moscow, ID 83844–3051 USA

I bruised my shin th' other day with playing at sword and dagger with a master of fence; three veneys for a dish of stewed prunes; and by my troth, I cannot abide the smell of hot meat since.

Shakespeare, Merry Wives of Windsor

Fight on, my men, Sir Andrew says. A little I'm hurt, but not yet slain; I'll lie down and bleed awhile, and then I'll rise and fight again.

Unknown, Ballad of Sir Andrew Barton

There is no dispute that ungulate sparring is somehow less serious than fighting; that it does not constitute a life or death battle. Sparring resembles a fencing match or cooperative dance in which rules are followed. The game is over after three 'veneys', or hits, have been made and the wager is not blood but rather a 'dish of stewed prunes'. Sparring appears to have, at the very least, a mild energy and small risk of injury cost. Such a behavior, if functionless, would have been weeded out long ago and would not be so popular an activity among a wide range of mammals today. What is the wager – spoils gained or lost – in a sparring match between two young ungulate males? To answer this question, we describe the context in which sparring is observed as well as the common participants in both pronghorn and other species in which sparring has been studied. In this and other recent studies, there is an attempt to quantify, rather than simply describe, sparring in a manner that will help to resolve its function(s) (Berger 1980; Byers 1980; Gomendio 1988; Schwede et al. 1990; Barrette & Vandal 1990; Rothstein & Griswold 1991; Hass & Jenni 1993; Watson 1993). First, we define sparring and describe the form that it takes in various organisms. We then present a detailed analysis of sparring in young pronghorn males.

What is sparring?

Sparring is a common type of social interaction observed in eutherian mammals (Fagen 1981) and in a few marsupials (Watson 1993). Sparring appears structurally similar to fighting, in whatever form that takes in a given species. In pronghorn it is head to head contact involving twisting and pushing with horns. However, sparring can include the orientation of other important weapons such as teeth, tusks, neck and forelegs (Walther 1984). Camels, llamas and horses often spar by rearing up on their hind legs, throwing their bodies at the opponent and simultaneously attempting to bite one another (Waring 1983; Berger 1986). In llamas, spitting is also common (Walther 1984). Rearing up is sometimes observed in deer, especially antlerless females and stags that have shed or have antlers yet in velvet (Walther 1984; Clutton-Brock et al. 1982). Necking in giraffes, or the rubbing of one male's head or neck against the neck of another, is observed both in sparring and fighting. In wallabies and kangaroos, sparring involves boxing and wrestling (Croft & Snaith 1991, Watson 1993).

Sparring rarely results in injury. Movements in sparring are slower and more gentle than those used in fighting, and communicative postures are different (Fig. 7.1(*a*)). As in forms of social play, a set of cues dictate the beginning and ending of sparring matches. The lowering of horns by one male is enough to initiate such a match (Fig. 7.1(*b*)), while the turning away indicates that it is over and the interaction is mutually terminated. Male pronghorn, at the termination of such a bout, occasionally swing their horns around and redirect sparring at the nearest bush. As in sparring matches observed in other organisms (Barrette & Vandal 1990; Schwede et al. 1990), pronghorn do not pursue or force an interaction after the opponent turns away. Sparring resembles a form of cooperation observed in play fighting (Fagen 1981).

Fighting, in contrast to sparring, poses a considerable risk of injury. Clerck (1965) observed a fight between two giraffes that left one individual with a spinal cord pierced by splintered neck vertebrae. Equids fight by biting at vulnerable tendons of the lower leg. Berger (1986) concluded that the teeth of equids are just as dangerous a weapon as any other. In red kangaroos, fighting involves powerful kicks to the abdomen with both hind legs and was observed by Frith & Calaby (1969) to cause one death by internal hemorrhaging. Clutton-Brock et al. (1984) esti-

(a)

(b)

Fig. 7.1. (a) A yearling male approaches another, inducing it to stand. (b) The two begin to spar. Note that the ears of both males are held forward; in broadside threats and in fights, the ears are held back.

mated that most red deer stags over five years of age averaged approximately five fights during the breeding season. Fights were generally preceded by roaring which, if the intruding male continued to approach, escalated into parallel walking and finally into fighting that could result in serious injury. Clutton-Brock et al. estimated that 23% of red deer stags over the age of 5 years were injured during the rut each year, while 6% were permanently injured. Common injuries included damage to the eyes (including blindness), injuries to the legs and major antler breakage. At the end of a fight, the winning stag would pursue the loser a short distance of 10–20 m (Clutton-Brock et. al 1984). Fighting in pronghorn is typically preceded by parallel walking, and only between closely matched individuals. Younger, smaller males do not parallel walk and do not fight with larger males. Cooperation is not observed in fighting; if an individual slips its rival immediately attempts to gore him. Byers (1997) reported that a pronghorn male experiences a 12% risk of death in a fight.

Serious fighting that results in injury occasionally occurs in juveniles. In pigs, fighting during the establishment of a 'teat order' is observed within 48 h of birth (Pond & Houpt 1978). Piglets are born with incisors and canines known as 'needle teeth' that they use to slash litter mates while competing for access to teats. Play fighting, the most common form of play in piglets, involves head to head combat and is characterized by the inhibition of biting (Pond & Houpt 1978). Fights among juveniles for milk also occurs in collared peccaries (Byers 1983). In spotted hyenas, highly aggressive newborns may kill one another (Frank et al. 1991).

Sparring is more common in male than in female juveniles (Wells & Goldschmidt-Rothschild 1979; Byers 1980; Fagen 1981; Gomendio 1988; Rothstein & Griswold 1991). In horses, male foals and yearlings occasionally play fight with adult males (Waring 1983). In giraffes, sparring or necking is seen only in males and includes adults, sub adults and calves (Dagg & Foster 1976). In bighorn ewes, head to head contact is rare, and usually vicious (Simmons 1980). Head to head contact is sometimes seen in female pronghorn, but only as the brief escalated end of a dominance interaction (Byers 1997). In red deer, sparring typically occurs between young stags, and as with pronghorn, never involves harem holding males (Clutton-Brock et al. 1984). In red deer the frequency of sparring declines after the fifth year (Clutton-Brock et al. 1984). Pronghorn males perform most sparring when one and two years of age while in bachelor groups. At ages three and older, males are increasingly likely to become solitary and site faithful in the summer (Byers 1997).

In pronghorn and many other ungulates sparring is most common in all male bachelor groups. Male pronghorn often exhibit locomotor–rotational movements prior to or immediately following sparring. This behavior is commonly associated with ungulate play and includes head shaking and rotation of the head and neck (Byers 1984). Sparring in pronghorn is separate from, and not an escalation of, dominance interactions such as displacements (see description below). In giraffes sparring was followed by resting, foraging, and sometimes male–male mounting (Dagg & Foster 1976). Mutual grooming is commonly observed between bouts of sparring (Waring 1983). Shoen et al. (1976) observed that 28% of play fights between foals were interrupted by facial gestures resembling flehmen. When approaching adult males for sparring, foals often perform a snapping display or the up and down movement of the jaw with the lips retracted (Waring 1983).

Hypotheses about the function of sparring

A list of proposed functions for sparring observed in various organisms is given in Table 7.1. In some cases, sparring is believed to serve a different function depending on the age of the participants. Sparring could function in; (1) the formation of a dominance order that influences current or future competition for resources; (2) motor training for future interactions; and/or (3) the development of social skills including assessment of other's abilities. One additional proposed function for social interactions, including play fighting, is 'social cohesion' (Bekoff & Byers 1981, Fagen 1981, Byers 1984, Pellis 1993). In highly social organisms, such as collared peccaries, group play may be a mechanism to maintain social bonds.

Because sparring resembles fighting, it is often assumed to have the same function: competition over resources or position in a dominance order. In contrast, play fighting, one of the most common forms of play, does not settle conflict over access to resources or result in a change in the dominance order (Fagen 1981). Motor training refers to the developmental effects of exercise, including cardiovascular and respiratory efficiency, enhanced muscle growth, and neuromuscular coordination (Bekoff & Byers 1981; Byers 1984; Byers & Walker 1995). Development of social skills through experience gained in interacting with conspecifics may be an important factor in determining an individual's lifetime reproductive success. This effect is most pronounced in young mammals where there is evidence that differences in early social behavior lead to differences in adult social behavior (Scott et al. 1974; Bekoff 1977, Fagen 1981). One

Table 7.1. *Proposed functions of sparring and/or play fighting.*

Proposed Function	Organism	References
Motor training and development of social skill in yearlings and adults	Pronghorn	This study
Motor training and development of physical strength and skill in juveniles	Siberian ibex, Bison, Bighorn sheep, Red necked wallaby	Byers 1980, Rothstein and Griswold 1999, Hass and Jenni 1993, Watson 1993
Immediate benefit to juveniles from information acquired about the new social environment	Cuvier's gazelle	Gomendio 1988
Tactile assessment of weapons and fighting ability in adults	White-tailed deer	Schwede et al. 1990
Assessment of fighting skills	Red kangaroo	Croft and Snaith 1991
Establishment of dominance	Giraffe	Dagg and Foster 1976
Socio-sexual bonding	Giraffe	Coe 1967; Geist 1966
Friendship maintenance in adults	Norway rat, Collared peccary	Pellis et al. 1993; Byers 1984

proposed function of sparring in adults is the assessment and comparison of weapon size and fighting ability prior to fighting (Table 7.1).

Variables used to quantify sparring include timing (both developmentally and seasonally), location (breeding or feeding grounds), and the occurrence of other activities such as play, aggression and sex during sparring behavior. In addition, partner preference and initiation and termination rates are frequently used in an attempt to resolve alternative functional hypotheses (Barrette & Vandal 1990).

Sparring in pronghorn males

Sparring is common in pronghorn males. Summer male, or bachelor, groups comprise males that differ in age and size. These easily observed groups provide an ideal setting in which to test the hypotheses concerning the function of sparring. In pronghorn bachelor groups, dominance interactions occur separately from sparring and result in a dominance hierarchy. Thus, by quantifying interactions between pronghorn bachelors, we sought to determine if sparring is a simple extension of dominance or if it is something else.

Methods

We collected data from 5 June to 22 August 1990 at the National Bison Range (Moiese, Montana, U.S.A.) a wildlife refuge encompassing 7,504 ha. The refuge is divided into pastures with fences raised to allow free movement of all ungulates except bison. A description of the refuge can be found in Byers (1997). We observed a bachelor group on Antelope Ridge located in Alexander Basin, an area in the North East section of the refuge. Roads allowed access to several vantage points from which the bachelor group was viewed.

This bachelor group of six bucks consisted of three adults and three yearlings (KA, four years; SS, three years; JR, two years; VO, LA, YV, yearlings). We distinguished individuals by characteristic neck bands and horns. Two of the individuals were tagged as fawns using plastic ear tags with a right–left color combination (Miller & Byers 1991). The group was observed from a vehicle or on foot at a distance of 200 m or more. Pronghorn at the Bison Range are habituated to the presence of vehicles and do not seem disturbed by the presence of a stationary observer at these distances. We sampled the behaviour of bachelors using a binocular and a 15×60 m spotting scope. Not all bachelors were present during all observation periods; sample size ranged from four to six individuals. We used all occurrence sampling (Altmann 1974) to record all sparring and dominance interactions during morning or evening observation periods of 1–4 h for a total of 150 bachelor hours of observation. Each interaction was classified as either a dominance displacement or a sparring match, and we recorded the initiator, recipient, and winner for all interactions observed.

Displacements were aggressive interactions used to assert dominance. This type of interaction occurred when one animal moved towards a second, that moved away, yielding its position. Displacements usually involved little contact, the initiator nudging or butting the rear or side of the recipient. The initiator of the displacement was defined as the winner if he was successful in displacing the recipient; the recipient of the interaction was defined as the winner if the initiator failed to displace him. A bedding displacement occurred when one individual approached another that was reclined, forcing it to stand and move away. A feeding

Table 7.2a. *Dominance hierarchy based upon displacement interactions.*

	Loser					
Winner	KA	SS	JR	VO	LA	YV
KA	–	3	7	8	8	3
SS		–	12	2	5	1
JR			–	12	12	4
VO		1	1	–	34	27
LA				7	–	37
YV				4	1	–

* For key see text

displacement occurred when a feeding individual approached another feeding individual who then moved away. A simple displacement occurred when interactants were standing or walking and not feeding.

We defined a sparring match as any head-to-head contact involving horn contact with head twisting and pushing. The initiator of the sparring match was the first individual to lower its head. A sparring match ended when one individual (the loser) turned away from its opponent (the winner). When both animals simultaneously switched to another activity, the match was considered a draw. For each sparring match, we recorded the duration and sparring intensity. We subjectively assessed the sparring intensity on a 10 point scale, where one was defined as touching horns and ten as all-out fighting.

Results

We observed 324 interactions. Of these, 182 were displacements (86 simple, 67 feeding, 29 bedding) and 142 were sparring matches. Yearlings initiated 63% of sparring matches and 59% of all displacements observed.

Dyadic dominance relationships were stable and dominance was transitive significantly more often than expected by chance (displacements: $N = 6, \chi^2 = 52.00, df = 30; P < 0.01$ sparring matches: $N = 6, \chi^2 = 63.00, df = 30, P < 0.005$; method given by Appleby 1983). Dominance hierarchies, constructed from the outcomes of displacements and sparring matches by minimizing the number of interactions below the diagonal, were identical and correlated with age (Table 7.2 a,b). The dominance hierarchy constructed from the outcomes of displacements was significantly more linear than expected by chance, but because of

Table 7.2b. *Dominance hierarchy based upon sparring interactions.*

	Loser					
Winner	KA	SS	JR	VO	LA	YV
KA	–	1	10	2	4	1
SS		–	6	10	2	*
JR	1		–	9	7	*
VO	1	1	2	–	23	25
LA				4	–	18
YV						–

* No interactions observed for dyad

two missing relationships, the sparring hierarchy was not (displacements: $K = 1, P < 0.022$ sparring $K = 0.66, P > 0.05$ Appleby 1983). We classified a winner and loser for all displacements, and for most sparring matches (nine sparring matches were draws).

For yearlings, the rate of displacements ($N = 54$; $X = 1.59/\text{h} \pm 0.19$) was nearly twice that of sparring matches ($X = 0.80 \pm 0.16$) In adults, there was a similar trend with displacements ($N = 43$; $X = 0.94 \pm 0.18$) occurring more frequently than sparring matches ($X = 0.65 \pm 0.11$) Overall, yearlings had significantly higher sparring rates (Nested ANOVA $F = 8.21, df = 1, 4, P < 0.05$) and displacements (Nested ANOVA $F = 9.14, df = 1, 4, P < 0.05$) than adults.

The rate at which each individual initiated and received sparring and displacements is shown in Fig. 7.2. The difference between initiation and recipient rates of displacements varied positively with dominance ($N = 6$, adj.$r^2 = 0.59$, $F1, 4 = 8.24$, $P = 0.045$) with more dominant individuals initiating more often than receiving. The difference in sparring, however, was not significantly different from a slope of zero ($N = 6$, adj.$r^2 = 0.42$, $F1, 4 = 4.67$, $P = 0.097$) The absolute difference between initiation and recipient rates was greater for displacements than sparring matches (Two way ANOVA, $F = 11.52, df = 1, 182, P = 0.001$).

Partner preference

To determine if each male exhibited social partner preferences, we compared the observed and expected frequencies of interactions that he initiated with other males. For each male, we calculated the expected frequencies of interactions with other males by dividing the number of interactions that the male initiated by the number of other males present.

Table 7.3. *Preferences of individuals for adult or yearling partners when initiating displacement and sparring interactions.*

	Adult		Yearling	
Individual	Expected	Observed	Expected	Observed
Displacements				
KA	8.9	8	18.1	19
SS	7.0	12	13.0	8
JR	9.3	2	17.8	25
VO	32.4	3	33.6	63
LA	20.2	1	16.8	36
YV	2.5	1	2.5	4
Sparring				
KA	6.6	11	11.4	7
SS	6.0	8	13.0	11
JR	5.8	2	10.3	14
VO	22.4	8	21.6	36
LA	14.5	1	13.5	27
YV	7.6	0	9.4	17

We compared the observed frequencies with those predicted by the null hypothesis of no partner preference. The results for each individual are summarized in Table 7.3.

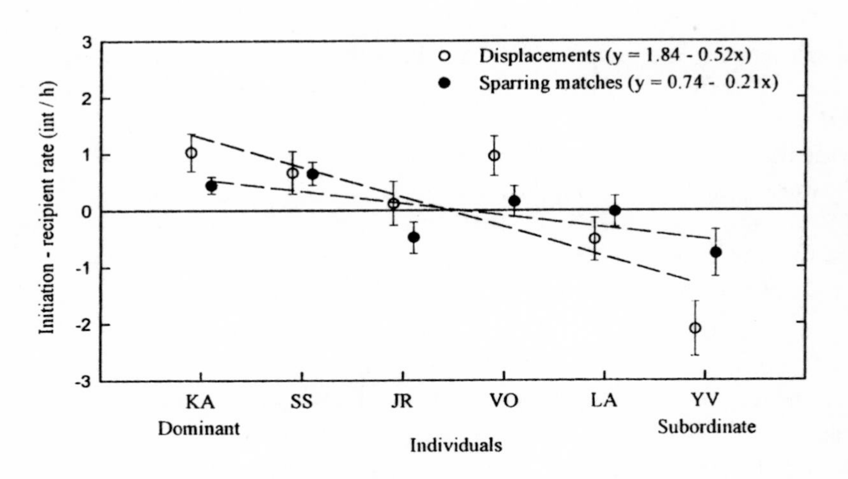

Fig. 7.2. Difference between initiation and recipient rates for displacements and sparring in relation to dominance. Error bars show the standard error.

Age and preference

For each individual, we analyzed partner preference at the level of preference for adults versus yearlings. Because of small expected values for some dyads, partner preference could not be further analyzed for preference for individuals. The proportion of both displacement and sparring match initiation by each age class is shown in Fig. 7.3. Yearlings preferred yearlings as social partners for both displacements and sparring matches. Only 6% of the total sparring matches were initiated by yearlings on adult males and all of these were initiated by the dominant yearling. For adults, there was no significant difference between the number of interactions observed and expected between adults and yearlings. However, the dominant adult (KA) initiated 11 of the 18 sparring matches observed with other adults and the subordinate adult (JR) initiated 14 of the 16 sparring matches with yearlings.

We ranked individuals from the dominance hierarchies and grouped individuals that interacted as being either one, two, three or more dominance ranks apart. Of the nine sparring matches ending in draws, four were between individuals differing in rank by two. Yearlings showed a

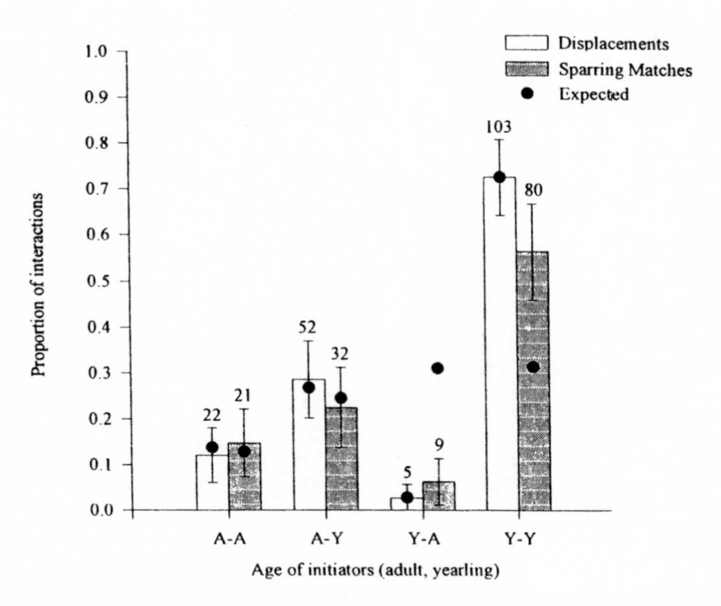

Fig. 7.3. The proportion of displacements and sparring matches initiated by adults (A) and yearlings (Y). Error bars are Bonferroni Z confidence intervals.

significant preference for males close in rank in both displacements (Fig. 7.4(*a*)) and sparring (Fig. 7.4(*b*)). In contrast, adults initiated significantly more sparring matches (Fig. 7.4(*b*)) than was expected with individuals that differed in rank by two. There was no difference in the expected and observed number of displacements initiated by adult males (Fig. 7.4(*a*)).

Dominance and rank

We grouped each interaction according to whether it was initiated by the dominant or subordinate individual of that particular dyad. Exactly 50% of all displacements were initiated by dominant individuals on subordinates that were only one rank apart (Fig. 7.5(a)). A similar preference was observed in sparring, in which 32% of the matches were initiated by the dominant individual on another close in rank (Fig. 7.5(*b*)). In sparring, however, subordinate individuals initiated as many interactions as expected with individuals that differed by one and even two ranks. This was not observed in displacements, where only 9% of the total interactions were initiated by the subordinate member of a dyad.

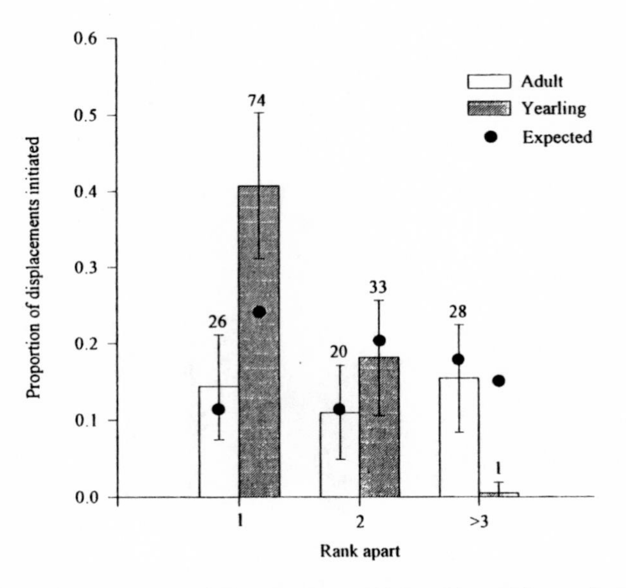

Fig. 7.4 (*a*). The proportion of displacements initiated by adults and yearlings on individuals of differing dominance rank. Error bars are Bonferroni Z confidence intervals.

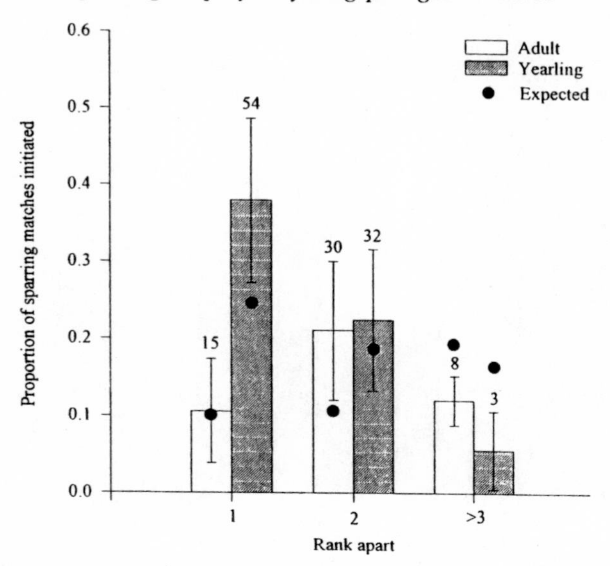

Fig. 7.4 (*b*). The proportion of sparring matches initiated by adults and yearlings on individuals of differing dominance rank. Error bars are Bonferroni *Z* confidence intervals.

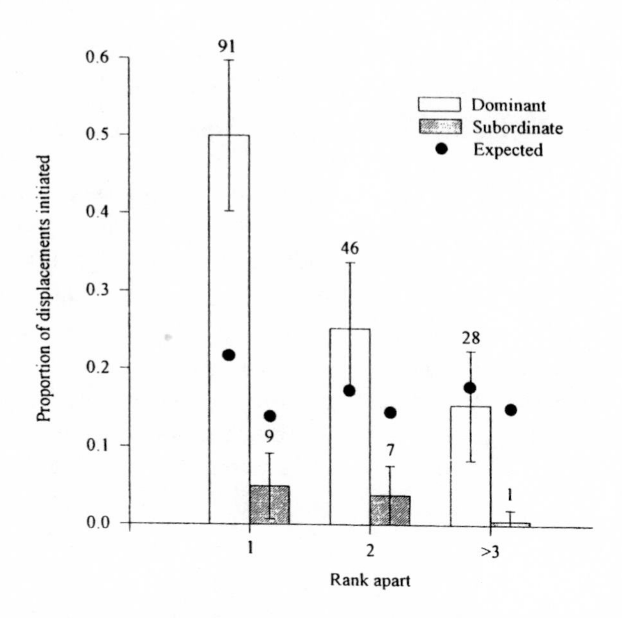

Fig. 7.5 (*a*). The proportion of displacements initiated by dominant versus subordinate males on individuals of differing dominance rank. Error bars are Bonferroni *Z* confidence intervals.

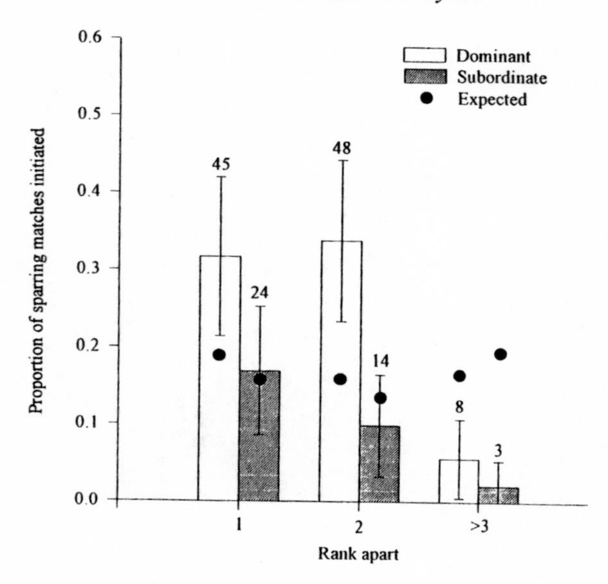

Fig. 7.5 (*b*). The proportion of sparring matches initiated by dominant versus subordinate males on individuals of differing dominance rank. Error bars are Bonferroni *Z* confidence intervals.

Sparring intensity and duration

There was no difference in the duration of sparring matches initiated by yearlings ($N = 87, X = 35.84 \pm 6.0$) and adults ($N = 52, X = 27.87 \pm 6.18$) Similarly, the intensity of sparring was not correlated with the age of the initiator. The average was relatively low on the scale of 1–10 for both yearlings ($N = 80, X = 3.11 \pm 0.19$) and adults ($N = 51, X = 2.65 \pm 0.21$) Duration and intensity of sparring was significantly higher for spar partners of the same versus differing age group (one-tailed Mann-Whitney tests, Duration: $U = 1574, P = 0.04$; Intensity: $U = 1335, P = 0.02$ Fig. 7.6). In sparring matches involving adults and yearlings, there was no significant difference in who initiated and duration ($U = 90.00, P = 0.42$) or intensity ($U = 74.5, P = 0.89$).

Rank

How far apart two individuals were in rank did not affect the duration (one-tailed Kruskall-Wallace $N = 139, \chi^2 = 3.81, P = 0.07$) or intensity (one-tailed Kruskall Wallace $N = 131, \chi^2 = 1.64, P = 0.22$) of the spar-

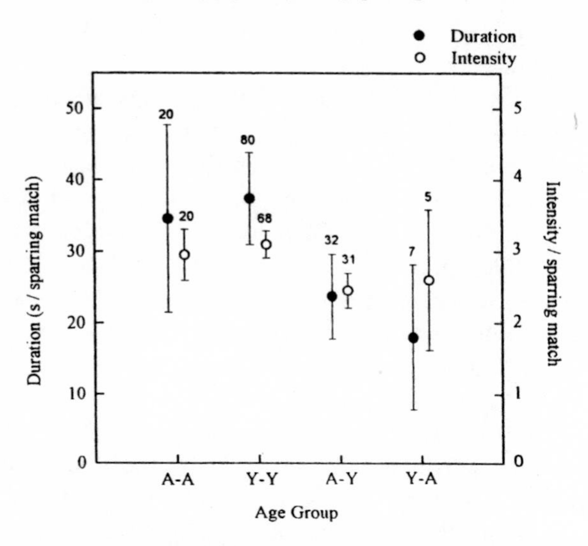

Fig. 7.6. The duration and intensity of sparring matches between individuals of differing age groups. (adults (A) and yearlings (Y)) Error bars show the standard error.

ring match. Who initiated, dominant or subordinate, did not significantly affect the duration or intensity of matches between individuals differing in rank (Mann-Whitney U, rank difference $= 1 : U = 383.5, P = 0.08$ rank difference $= 2 : U = 284, P = 0.62$ rank difference $= 3 : U = 6.0, P = 0.59$) or intensity (rank difference $= 1 : U = 210, P = 0.85$; rank difference $= 2 : U = 288.5, P = 0.67$; rank difference $= 3 : U = 53.5, P = 0.29$).

Discussion

Our results support the hypothesis that sparring is a cooperative activity in which males improve some of the motor skills involved in fighting. The prediction that individuals should choose closely matched sparring partners was clearly demonstrated for yearling pronghorn. A comparison with displacements showed that yearlings preferred other yearlings and had higher rates of both displacing and sparring than adults. Rothstein & Griswold (1991) found this type of preference in bison juvenile social interactions. Because young animals are just becoming established in a dominance order, it is predictable that they would also have higher rates of displacing than adults.

Among adults, preference for social partners was mixed, suggesting that these types of interactions among bachelors are not as crucial to adult fitness and therefore partner preference is more flexible; adults stand to benefit less from motor training than yearlings and have already established their dominance order. The identical dominance hierarchies constructed from the outcomes of sparring and displacements could be viewed as evidence for the social dominance hypothesis. Unlike bighorns, where dominance relationships in yearlings was vague (Hass & Jenni 1991), dominance among pronghorn yearlings was clearly established by displacements.

Dominance affected sparring behavior. Nearly half of all sparring matches initiated by yearlings were initiated by the dominant yearling (VO), while the most subordinate yearling (YV) initiated less than one out of five sparring matches observed between yearlings. In addition, the dominant yearling sparred most frequently with adult males.

That dominance influenced the rate and preference with which pronghorn males initiated sparring matches could be seen as support for the hypothesis that males use sparring to establish and reinforce dominance relationships. However, the effect of dominance on sparring could be a byproduct of a hierarchy already established through displacements rather than a mechanism by which dominance is initially settled. As predicted by Fagen (1981) the social hierarchy in which organisms develop can profoundly affect the pathways the individuals take. In this study, the social environment allowed the dominant yearling to interact more frequently with older bucks and to perhaps learn successful strategies more readily with older more skilled partners (Fagen 1981). Although fighting ability is generally considered to be correlated with reproductive success (Clutton-Brock et. al. 1984), males that lack experience provided through sparring may opt for other reproductive strategies. One pronghorn male, the subordinate member of a bachelor group in 1984, successfully defended a group of females by hiding them in an area of the bison range not frequented by other reproductive males, hence avoiding any confrontations with similar sized males.

Sparring is most often observed in males and this sex difference has been used to support the motor training and social skills hypothesis for the function of sparring. Both of these hypotheses predict that males are gaining either social experience or coordination that they will use in more serious interactions such as courtship or fighting. Gomendio (1988) argued, however, that the benefits of play fighting need not be delayed but rather may allow individuals to gain immediately pertinent information about the new social environment. The rate of play fighting in captive Cuvier's gazelles peaked at four months of age and began to decrease considerably around six months. If play fighting was just motor training for future behavior, Gomendio (1988) reasoned, then one would not expect to a see a sudden drop in its occurrence. However, Gomendio's study did not follow individuals past seven months and sparring, other than that between the young juveniles and adults females and males, was not described for older individuals. Miller & Byers (1991) found that locomotor play decreased during weaning and Gomendio's (1988) results may suggest a similar trend.

Intensity of sparring and partner preferences in sparring are commonly offered as evidence about the function of sparring. This emphasis, however, is problematic since the predicted outcomes for the hypothesized functions are similar, especially those for motor training and social dominance. To obtain the maximum benefit from training, individuals should prefer to spar with those of the same sex, age and skill. Unfortunately, those individuals are also most likely to be closest in dominance and should be preferred during the establishment and reinforcement of a hierarchy.

Barrette & Vandal (1990) proposed a way to distinguish between the alternating hypotheses for the function of sparring by examining the initiation and termination of bouts between mismatched pairs of adult male caribou. If sparring functions as training, they predicted that all individuals would initiate and terminate sparring bouts equally regardless of their antler size. However, if sparring establishes dominance, they predicted that smaller males would initiate less and terminate more of the sparring bouts. Males that use sparring for assessment of size and fighting ability, are predicted to initiate sparring equally but smaller males, upon determining their opponent's ability, would terminate more than larger males. In their study, male woodland caribou were ranked according to size of antlers, a measurement which did not necessarily correspond to their age or weight. They concluded that since the smaller males in their study initiated roughly 50% of the time, while terminating those bouts 90% of the time, the function of sparring was most likely the assessment of fighting ability.

In contrast to Barrette & Vandal's result, Wachtel et al. (1978) found that in mule deer, bucks with larger antlers initiated more frequently while the smaller males were more likely to terminate the interaction.

This finding suggests, according to the criteria established by Barrette & Vandal (1990), that sparring in mule deer during the rut functions in the maintenance of a dominance hierarchy. In red deer, sparring bouts were usually initiated by the dominant member of the pair and generally terminated by the subordinate (Clutton-Brock et al. 1984).

In pronghorn we found that the dominance hierarchy, based on who terminated or lost the interaction, was correlated with age and size of males. This dominance hierarchy provided a basis for ranking pronghorn dyads. In dyads where males differed in rank by one or two, dominant males initiated a greater proportion of sparring matches than subordinate males (Figure 7.5(*b*)). According to Barrette & Vandal's (1990) criteria, these data suggest that the function of sparring is the maintenance of a dominance hierarchy. However, adult males initiated more sparring matches than was expected with males that differed in rank by two (Figure 7.4(*b*)). This suggests, as does the context in which it is observed in the field, that sparring behavior is not an escalation of dominance interactions. The difference in initiation and recipient rates among males also suggests that sparring is more egalitarian than displacements; dominant males initiate more displacements than they receive, but there is no such trend in sparring (Fig. 7.2). In addition, the lack of a correlation between duration and intensity of sparring and dominance of sparring partners suggests that sparring is not used to settle contested hierarchies.

Coe (1967) believed that necking behavior in giraffe bachelor herds functioned to set up and maintain a hierarchy of sexual and physical dominance. However, in pronghorn, bachelor groups include yearlings that are obviously subordinate to adults and will not compete directly for access to females. In addition, the older bucks with higher reproductive success are not present in bachelor groups. Finally, the bachelor group is maintained over several months with the adult members coming and going.

In red deer, a male's fighting success played a major role in its reproductive success (Clutton-Brock et al. 1984). The reproductive success of harem holding males was correlated to the size of the harem and how long a male could hold on to it against intruding males. The importance of fighting ability in determining reproductive success, the frequency of sparring among young males, and the occurrence of other, less costly interactions used to establish and maintain dominance all suggest that sparring may function as training for future interactions. We cannot exclude the establishment and maintenance of a dominance hierarchy as a function of sparring, but dominance alone seems a poor explanation for the prevalence and egalitarian nature of this behavior in pronghorn bachelor groups.

Acknowledgments

Our work was supported by grants NICHD 22606 and National Geographic Society 4759–92 to J. A. Byers. The United States Fish and Wildlife Service permitted us unrestricted access to the National Bison Range. Bison Range personnel provided diverse cooperation, assistance, and favors.

References

Altmann, J. 1974. Observational study of behavior: sampling methods. *Behaviour*, **49**, 227–67.

Appleby, M. C. 1983. The probability of linearity in hierarchies. *Anim. Behav.*, **31**, 600–8.

Barrette, C. & Vandal, D. 1990. Sparring, relative antler size, and assessment in male caribou. *Behav. Ecol. Sociobiol.*, **26**, 383–7.

Bekoff, M. 1977. Mammalian dispersal and the ontogeny of individual behavioral phenotypes. *Am. Nat.*, **111**, 715–32.

Bekoff, M. & Byers, J. A. 1981. A critical reanalysis of the ontogeny and phylogeny of mammalian social and locomotor play: an ethological hornet's nest. In: *Behavioral development* (eds. K. Immelmann, G. Barlow, M. Main, and L. Petrinovich), pp. 296–337. Cambridge: Cambridge University Press.

Berger, J. 1980. The ecology, structure, and functions of social play in bighorn sheep. *J. Zool. Lond.*, **192**, 531–42.

Berger, J. 1980. Wild horses of the Great Basin: social competition and population size. Chicago: University of Chicago Press.

Byers, J. A. 1980. Play partner preferences in siberian ibex, *Capra ibex sibirica. Z. Tierpsychol.*, **53**, 23–40.

Byers, J. A. 1983. Social interactions of juvenile collared peccaries, (*Tayassu tajacu*). (Mammalia: Artiodactyla). *J. Zool. Lond.*, **201**: 83–96.

Byers, J. A. 1984. Play in ungulates. In: *Play in animals and humans* (ed. P. K. Smith), pp. 43–65. Oxford: Basil Blackwell.

Byers, J. A. 1997. *American pronghorn. Social adaptations and the ghosts of predators past.* Chicago: University of Chicago Press.

Byers, J. A. & Walker, C. 1995. Refining the motor training hypothesis for the evolution of play. *Am. Nat.*, **146**, 25–41.

Clerck, A. de. 1965. News from our parks. *African Wild Life*, **19**, 34.

Clutton-Brock, T. H., Guinness, & F. E., Albon, S. D. 1982. *Red deer. Behavior and ecology of two sexes.* Chicago: University of Chicago Press.

Coe, M. J. 1967. "Necking" behaviour in the giraffe. *J. Zool.* (London), **151**, 313–21.

Croft, D. B. & Snaith, F. 1991. Boxing in red kangaroos, *Macropus rufus*: aggression or play? *Int. J. Comp. Psych.*, **4**, 221–36.

Dagg, A. I. & Foster, J. B. 1976. *The giraffe: its biology, behavior, and ecology*. New York: Van Nostrand Reinhold Co.

Fagen, R. 1981. *Animal play behavior*. London: Oxford University Press.

Frith, H. J. & Calaby, J. H. 1969. *Kangaroos*. Melbourne: F. W. Cheshire.

Frank, L. G., Glickman, S. E., & Licht. P. 1991. Fatal sibling aggression, precocial development, and androgens in neonatal spotted hyenas. *Science* **252**: 702–4.

Geist, V. 1966. The evolution of horn-like organs. *Behaviour*, **27**, 175–214.

Gomendio, G. 1988. The development of different types of play in gazelles: implications for the nature and functions of play. *Anim. Behav.*, **36**, 825–36.

Hass, C.C. & Jenni, D. A. 1991. Structure and ontogeny of dominance relationships among bighorn rams. *Can. J. Zool.*, **69**, 471–6.

Hass, C.C. & Jenni, D. A. 1993. Social play among juvenile bighorn sheep: structure, development, and relationship to adult behavior. *Ethology*, **93**, 105–16.

Kitchen, D. W. 1974. Social behavior and ecology of the pronghorn. *Wild. Monogr.*, **38**, 1–96.

Miller, M. N. & Byers, J. A. 1991. Energetic cost of locomotor play in pronghorn fawns. *Anim. Behav.*, **41**, 1007–13.

Pellis, S. M., Pellis, V. C., & McKenna, M. M. 1993. Some subordinates are more equal than others: play fighting amongst adult subordinate male rats. *Aggressive Behav.*, **19**, 385–93.

Pond, W. G. & Houpt, K. A. 1978. *The biology of the pig*. Ithaca, New York: Cornell University Press.

Rothstein, A. & Griswold, J. G. 1991. Age and sex preferences for social partners by juvenile bison bulls, *Bison bison*. *Anim. Behav.*, **41**, 227–37.

Schwede, G., Holzenbein, S. & Hendrichs, H. 1990. Sparring in white-tailed deer (*Odocoileus virginianus*). *Z. Saugetierk.*, **55**, 331–9.

Scott, J. P., Stewart, J. M. & De Ghett, V. J. 1974. Critical periods in the organization of systems. *Dev. Psychobiol.*, **7**, 489–513.

Simmons, N. M. 1980. Behavior. In: *The desert bighorn: its life history, ecology, and management* (eds. G. Monson and L. Sumner), pp. 124–144. Tucson: University of Arizona Press.

Schoen, A. M. S., Banks, E. M. & Curtis, S. E. 1976. Behavior of young Shetland and Welsh ponies (*Equus caballus*). *Biol. Behav.*, **1**, 199–216.

Wachtel, M. A., Bekoff, M., & Fuenzalida, C. E. 1978. Sparring by mule deer during rutting: class participation, seasonal changes, and the nature of asymmetric contests. *Biol. Behav.*, **3**, 319–30.

Walther, F. R. 1984. *Communication and expression in hoofed mammals*. Bloomington: Indiana University Press.

Waring, G. H. 1983. *Horse behavior*. New Jersey: Noyes publications.

Watson, D. M. 1993. The play associations of the red-necked wallabies (*Macropus rufogriseus banksianus*) and relation to other social contexts. *Ethology*, **94**, 1–20.

Wells, S. M. & Goldschmidt-Rothschild, B. von. 1979. Social behavior and relationships in a herd of Camarague horses. *Z. Tierpsychol.*, **49**, 363–80.

8

Squirrel monkey playfighting: making the case for a cognitive training function for play

MAXEEN BIBEN
2014 Martinsburg Pike, Winchester, VA 22603 USA

Saimiri play in wild, semi-wild, and captive environments

Squirrel monkey (*Saimiri sciureus*) play behavior was first described almost 30 years ago in studies by Frank DuMond (1968) and John Baldwin (1969, 1971). These early reports were made under the benign observation conditions of the Monkey Jungle facility near Homestead, Florida. The provisioned four acre enclosed forest, referred to in the DuMond and Baldwin papers as a 'seminatural' environment, was home to over 100 squirrel monkeys of all ages, as well as other primate species. The extensive descriptions of play behaviors, partner preferences, and ontogeny observed at Monkey Jungle were confirmed a few years later by Baldwin's observations of wild squirrel monkeys in Panama and Costa Rica. At the National Institutes of Health 'animal farm' facility at Poolesville, Maryland, I observed monkeys in indoor and outdoor enclosures and found play indistinguishable from the earlier published descriptions. Only the frequency of play, responding to environmental stressors such as food shortages, was variable. It is reassuring that play has a basic conservatism and a reliability of behavior patterns over a wide range of settings, for much of our research has come from studies of captive animals.

At about five weeks of age, infants start to reach out to each other in rudimentary play while mounted on the backs of mothers who are resting side by side. During the second month, the infants slide off their mothers' backs, and wrestle vigorously beside the mothers as they rest, or try to rest. Nervous mothers retrieve very young infants from these earliest play fights. Later, it is rare for a mother to intervene, even if the play gets noisy. Mothers in squirrel monkeys and other primate species apparently perceive the difference between play fighting and serious fighting (Aldis 1975; Humphreys & Smith 1987; Pelligrini 1989; Pellis & Pellis 1996).

One of the most important variables governing the occurrence of play at any age is the availability of a playmate. As in most New World primates, other youngsters, not the mother, are the typical playmates. In wild squirrel monkey troops, births tend to be synchronous (Boinski 1987). Mothers stay close to one another at the time of parturition, ensuring a pool of like-aged play partners in addition to older youngsters. With peer playmates readily available, an input of time and energy by the mother (or other adults) is apparently neither needed nor provided.

The vigorous contact form of play known variously as rough and tumble, play fighting, or wrestling play predominates in squirrel monkeys, as in many other mammals, as the major means of peer social interaction in the young during most of their development. Locomotor play (running, jumping, and swinging from branches or vines) also occurs but it has a heavy social component and often culminates in a bout of play fighting when one individual catches up with the other.

Dominance: the power role in play

Infant squirrel monkey males, like infant human males, are statistically larger and heavier than females from birth. By ten months of age, the males in a group of five males and five females studied in our laboratory were 15–20% heavier than the females and consistently demonstrated dominance over them (Biben 1986). Using the reliable indicator 'head grab' as a measure of dominance in natural (non-test) social encounters, juvenile males maintained a linear dominance hierarchy over a period of at least six months, with no within-sex relationship between weight and dominance. The five juvenile females had a stable but nonlinear dominance hierarchy.

Why and how juvenile males achieve dominance differentials is not clear. The fact that they do achieve them, and that it is apparently very important to them, is evident in their play interactions. Typically, males wrestled on flat surfaces where they could effectively pin each other down, in a way that clearly differentiates dominant from subordinate (Panksepp et al. 1978; Symons 1978). During a bout of wrestling, males maneuver continually to be in the dominant, 'on top' position.

While the appearance of such wrestling bouts never lost the unmistakable give and take of play, close observation revealed that achieving the on top position was not random. Being socially dominant was more important than being heavier. In a wrestling bout, the individual that was dominant outside of play spent more time in the on top position than did the individual who was subordinate to him. The significance of this was not lost on youngsters: they preferred to play with individuals they could dominate and avoided those who dominated them. Whether or not play, or winning, has immediate benefits, squirrel monkeys play to win, and they stack the deck by choosing both same and opposite sex partners that they can dominate.

How dominance relationships determine play decisions

Exercising control over whom you play with can be done in two ways, by soliciting play from a desired partner or by accepting or refusing another's solicitation. Squirrel monkeys do both, using a few well-chosen (that is, statistically associated with play) behaviors to signal their intent.

Males directed about twice as many play initiation behaviors to other males as to females, and were about twice as successful in getting males to play with them, resulting in about three times as much male–male as male–female play (Biben 1986). In same-sex interactions, males were significantly more likely to target individuals who were subordinate or about equal in dominance to them. A similar tendency for partners to be closely matched has been observed in the rough and tumble play of young boys (Humphreys & Smith 1987; Boulton 1991).

Unlike males, females were equally successful in their attempts to initiate play with partners of either sex, but females made many fewer attempts than did males, and were more likely to target other females, particularly ones subordinate to themselves (Biben 1986). Partly as a consequence of these choices, play in this species, as in many other primates, is disproportionately sex-segregated, with male–male play being more frequent than female–female or male–female.

This rigid and joyless description of play fights may give the impression that the outcome of wrestling bouts is a foregone and dreary conclusion for females, and subordinates of either sex. It is not. If we look at the overall dominance relationship of two individuals outside of play, and then during play, we find that the identity of the dominant and subordinate remain the same *but* the extent of the difference in the dominance relationship is significantly diminished during play, in a reversal of roles that is more relative than absolute.

We can represent this numerically by counting the dominance displays between animals A and B *outside* of play (head grabs) as opposed to *during* play (pins and head grabs). Typically, two males A and B will have a rather skewed dominance relationship outside of play, with A, for instance, head grabbing B about 30% more than the other way around. During play, this might drop to less than 10%, significantly less, but with A still dominant to B. B's chances of having what we might refer to as a 'dominance experience' are greatly improved when *playing* with A than they would be otherwise. In fact, the small margin of dominance maintained by males during play is the least skewed of any dominance relationship observed in juvenile squirrel monkeys of either sex, whether during or outside of play. Females maintain polarized relationships of about 75% dominance difference outside of play, with role reversal during play bringing them down to about 45%. Given these odds of 'winning' or being dominant in a play encounter, squirrel monkeys showed that winning was important by voting with their feet and playing more often with those partners whom they could dominate. In squirrel monkeys, the reversal of roles is generally *inter*bout rather than *intra*bout: in most cases, one individual maintains dominance throughout a bout and assumes the subordinate role in a later bout.

Role reversal: playing at different power roles

What role reversal means in a squirrel monkey play fight is that if A is dominant to B, then A is (apparently) allowing B to assume the dominant role more often than B could do on his own. Such self-handicapping probably promotes play by diminishing discouraging differences in dominance. Subordinate youngsters get enough opportunities to be dominant to keep them interested while dominant ones generally retain just enough of an edge (Bekoff 1978). Choosing to play or not play with any individual is an option in most cases (unless the pool of potential partners is very small). Without role reversal, presumably, nobody would want to play with the most dominant male youngster in the group (Altmann 1962; Aldis 1975).

We usually speak of role reversal in contact play but its significance transcends this. In children's fantasy play (and in some forms of adult play like festivals or gambling, where an underclass may have a temporary delusion of power) Sutton-Smith (1993, 1995) has referred to 'inversion of power' as a very seductive theme. He presents the example of his student Diana Kelley-Byrne, who established a play relationship with a seven year old by letting the youngster choose whatever game she liked. The little girl immediately chose to make Kelley-Byrne her slave. We forget how powerless a child's normal situation is, and we may underestimate what a thrill it is to act out powerful roles in play. 'Empowerment' in the here and now may be a large part of what is experienced as 'fun', and as such may be one of the *immediate* benefits enjoyed by youngsters when they play. There may also be long term benefits to feeling empowered in play. Psychiatrist Stuart Brown (Chapter 12), reviewing the cases of several severely antisocial men, noted abject deficits in their play as children and speculated that play may be needed to develop empathy and to handle feelings of powerlessness.

Playing the more powerful role in 'Cowboys and Indians' type games, achieving dominance by winning at 'King of the hill' type games, winning at games of strategy, and winning in wrestling play are all perceived as being more 'fun' and a goal to be achieved. Sutton-Smith (1995) ventures only that the adaptive value of feeling empowered in play is the assurance that life itself has meaning. While we may speak comfortably about animals experiencing 'fun' (Fagen 1992), self-validation is more of a stretch. But why not invoke fun? After all, squirrel monkeys do not actually improve their dominance status in play (Biben 1986) and no resources are garnered thereby, so there is no immediate 'serious' benefit or incentive there.

When role reversal isn't granted: alternative ways to avoid being dominated

Role reversal goes a long way toward equalizing dominance relationships in both male–male and female–female play, but chivalry is dead when males play with females: role reversal is not extended to them. Small wonder that males are less successful in coaxing play from a female than from a male partner, and that females target other females for their play. In *Saimiri*, as in many other primates, females play less frequently than males. The poor odds they get from males and, indeed, the still very skewed dominance relationships they experience in wrestling play with other females, may have a chilling effect on females' play. In either case, it is easy to predict which interactant will spend the most time in the dominant role.

It is perhaps not surprising, then, that wrestling bouts with pins and overt displays of dominance are not particularly popular with females. Examination of the patterns used in *Saimiri*'s wrestling play reveals that there are actually two alternative strategies. Wrestling to a pin, typically performed on a flat surface where one can effectively hold the partner down (termed *directional* wrestling), was only one such strategy and it was strongly favored by males. Females, on the other hand, were more likely to wrestle by hanging from a branch or perch with their forelegs clasped about one another (*nondirectional* wrestling).

In nondirectional wrestling, no dominant and subordinate distinction is evident. This more egalitarian form was preferred by all females and also by those males who were unlikely to get a good deal during directional wrestling. It was more common when females were the initiators and males the targets than vice versa (Biben 1986).

Regardless of the gender of the interactants, the probability of dominating another individual was a determinative factor in the type of wrestling initiated. With a statistically significant likelihood, animals initiated more directional wrestling when they were likely to be winners and more nondirectional wrestling when they were likely to be losers at directional wrestling. When the dominance relationship was very skewed, with one individual being dominant more than 60% of the time, the frequency of play suffered considerably (Biben 1989).

In light of these results, we see that play is hardly random cavorting. Instead, it is decidedly goal-directed, with dominance as the goal: animals play to win; if they can't win (and females usually can't) they choose games that don't involve winning or clear displays of dominance. Boulton (1991) makes the important point that competing to win, characteristic of boys' rough and tumble as well as that of squirrel monkeys, is part of the competitiveness of play, and not to be considered competition for resources.

As in most polygynous primates, squirrel monkey males play more roughly and more often than females and persist in play behavior to a later age. Females, who become sexually mature at an earlier age, turn their attention to playing with infants, babysitting and infant care. Gender differences like these are generally attributed to male hormones, starting in the prenatal environmental (Cheney 1978). While hormones no doubt direct these differences, we can see in squirrel monkeys some of the more proximal ways that hormonal differences translate to playing by different rules, thus maintaining gender segregation:

(1) males play more than females *because they enjoy greater opportunities for role reversal and thus are more likely to respond to invitations to play*

(2) like-sexed partners are preferred *because females don't get a fair shake in males' play and don't want to play with them*
(3) males are more likely than females to engage in rough directional play *because their greater likelihood of role reversal makes such play more fun for them than it is for females*
(4) males' play persists longer in ontogeny than females' *because males are in fact learning something about fighting from play that females don't need to know as much about (more about this later!)*

The interaction between gender differences in play styles and partner preference is not unique to monkeys. Sex-segregated play is a robust, cross-cultural phenomenon in children, with same sex preferences for play partners evident by two years of age (Howes 1988). In cultures across the world, there is a consistent finding that boys, more than girls, engage in rough and tumble play (Whiting & Edwards 1973; Humphreys & Smith 1984). Research on the play of young children has shown that gender preferences for particular styles of play interactions, utilizing different toys and games, tends to reinforce sex segregation (Erwin 1993). Girls play in small groups or pairs, learning to relate to others on an individual basis, picking up subtle social cues, and using the group as a source of support. Boys play in larger groups, are more physically active, and have more rules, goals, and physical competitiveness. Boys show little tolerance for girls' more relaxed and informal relationships, and use their own groups as a springboard for learning to defy authority.

How important is play? winning? the type of experience?

The preceding descriptions of wrestling play in *Saimiri* make it appear to be a rather hazardous (psychologically if not physically) undertaking where it can be tough to have a good time. Dominance remains the major theme, with the evolution of strategies like role reversal and non-directional wrestling to mitigate its heavy hand, and imbue play with the unpredictability and chance to play at power roles that are perceived as fun. Such strategies are easily seen as ways to increase the frequency of play. The obvious question then, becomes 'Why?' Why is increasing the frequency of play important? And why does play revolve around dominance and wrestling, rather than some other less contentious theme?

Is the opportunity to play more important than winning?

While the concept of a play 'drive' is no longer seen as a relevant issue in the research literature, there is evidence to suggest that play is a priority and that youngsters (and sometimes the not-so-young) will alter their 'normal' or expected behavior in order to have play experiences. Such evidence, while indirect, is relevant because more direct information, as from deprivation studies, is notoriously difficult to obtain for play.

Deprivation experiments are more often than not inconclusive: those that succeed in depressing play also depress other social behavior (Bekoff 1976; Fagen 1981; Martin & Caro 1985; Thor & Holloway 1984). As an alternative, some studies have sought to alter, rather than eliminate, playful behavior, e.g. by providing only atypical or nonplaying partners (Einon et al. 1978). In a mixed-age and -sex social group, squirrel monkeys choose partners and games that are advantageous to themselves. Young animals of other species do the same (Byers 1980; Fagen 1984). What happens if preferred partners are not available? Do animals opt for *no* play experience over possibly disadvantageous experiences? In primate species, youngsters lacking peers turn to solitary play or play with objects (Goodall 1986, 1990; Jolly 1985; Walters 1987). Another alternative, used by *Saimiri*, is to play with adults, who can provide a youngster with at least some of the same experiences that appear to be important in peer play, namely experience in the on top role during wrestling. Would youngsters be as 'generous'?

In a captive situation, it is possible to manipulate a youngster's social milieu to answer this question. Mindful of the dominance differentials characteristic of males and females (males always dominant to females) and males of different ages (older dominant to younger) and the preference for like-aged and like-sex play partners, we confined year-old male squirrel monkeys, each to a cage with two companions at a time. Sometimes the companions were of the 'preferred' variety (other year-old males) and sometimes not. Nonpreferred companions were older (two-year old) males and one-year old and two-year old females. Companions could play with each other or with the year-old males. What we found was that animals opted to play, at a rate equivalent to that seen in a free choice social situation, regardless of the type of companions to which they were restricted.

What changed was the way in which these males played. In a free-choice situation, males tend not to reverse roles in directional wrestling with females, but here they did. The two-year old male companions were not so generous, however: they consistently dominated the one-year olds, and role reversed only grudgingly. So for them, males ratcheted down the stakes by switching from directional to nondirectional play.

The unusually high use of role reversal between males and females and nondirectional wrestling between males can be seen as dual strategies promoting the likelihood of play. By being flexible, animals maintained their levels of play at about normal rates, despite the fact that, given a larger choice of playmates, these same animals would be unlikely to play together.

The resiliency of squirrel monkeys in accepting less than ideal playmates does have limits. Children's play frequency reflects their degree of liking for each other (Boulton 1991); similarly, squirrel monkeys confined with only a single partner as a potential playmate played more with some than with others (Biben & Champoux 1996). In this last study, youngsters adjusted, or didn't, to new partners within a few weeks, as indicated by changes in their plasma cortisol levels. Frequency of play was negatively correlated with cortisol levels. However, after being confined with each other for several months, animals that were uncomfortable in each other's presence showed a frequency of play that belied their level of stress.

Further evidence of the importance of play and the way in which role reversal is used as an enticement comes from our observations on play between parents and youngsters (Biben & Suomi 1993). As is typical in the New World primates (in fact, all nonhuman primates except the Great Apes), play is an activity *of* the young *with* the young. Even tiny infant squirrel monkeys on their mothers' backs play with *each other*. While adults, particularly in captivity, do occasionally play with each other, at no point in ontogeny is one likely to see an adult, related or otherwise, playing with a youngster. However, this statement only holds true if there are youngsters around with whom to play. In captivity, this is not always the case, and squirrel monkeys appear not to have solitary forms of play in their repertoire. Even in an extreme case where an infant was hand-reared in our laboratory out of social contact with other squirrel monkeys, the youngster treated his own tail as a social companion, grabbing it and wrestling with it while emitting play peeps, in a manner more reminiscent of social play than object play.

When an infant has no other youngsters to play with, it is a good bet that an adult will respond to the infant's attempts to initiate play. We found adult–infant play to be a virtual certainty in groups with only one infant, with the likelihood of such play dropping precipitously as the number of infants increased. In our lab, that adult was usually the dominant (and only) adult male in the group. If no adult male was present in the group, then an adult female was the next most likely partner, but we never observed the mother playing this role. With a size mismatch like that seen in youngster/adult play, self-handicapping was in order, and it is quite amusing to see a fully adult male weighing over 1 kg going to outrageous lengths in assuming the subordinate role, lying supine, flailing his limbs, exposing his fully adult canines in an enormous open mouthed grin, while a 300 gm youngster repeatedly leaps upon and attacks him. There are no reports of such behavior from adults in groups with multiple youngsters.

Cost/benefit ratio of squirrel monkey play fighting

The value of play is often questioned. While the persistence of *Saimiri* play under a variety of less-than-ideal circumstances is in itself evidence of its value, additional confirmation comes from the extent of risk incurred through play activity. Three types of risk appear relevant as potential costs in squirrel monkey play: physical risk (see Fagen 1995 for references on morbidity and mortality associated with play), energy expenditure, and social risk.

Physical risks

In children, rough and tumble has often been considered a form of aggression rather than play, by researchers, parents, and caretakers (Sutton-Smith 1993; Pelligrini 1995a). Observers of animal play have described play fights breaking down into serious fighting, and it is commonly believed that play fighting is practice for serious fighting, harmless only because the youngsters do not have the physical capabilities to hurt each other. Yet, serious fighting often coexists in the repertoire with play fighting and is readily differentiated from it by a variety of cues (e.g., Bekoff 1995; Pellis & Pellis 1996).

The breakdown of play into aggression actually seems to be rare in animals; although the threat of its occurrence might well have a significant effect on behavior. I have never observed an injury to occur to a youngster during play fighting. Nor have I ever seen a play fight become a serious fight, even among juveniles or subadults, who fight more often than younger animals. Two-year old squirrel monkeys, still at the height of playfulness, go out of their way to make mismatched partners comfortable in play (Biben 1989). Slightly older squirrel monkeys (2–3 years) play fairly or don't play at all, when their choice of playmates is restricted (Biben & Champoux 1996). Age does figure as a significant variable in boys: adolescents, especially rejected boys, are more likely than younger boys to be aggressive during rough and tumble (Pelligrini 1995b).

If play is distinct from fighting, and youngsters don't hurt each other in play fighting, is squirrel monkey play then safe? Interestingly, an easy way to distinguish squirrel monkey play and fighting is the presence of a distinctive vocal accompaniment to play. These vocalizations identify to an observer the 'safe' from the hazardous activity, but the noise may add an element of hazard to play. Youngsters at play are preoccupied and vulnerable. Noisiness would not only make it that much more difficult for them to hear danger coming or to respond to another's alarm call, it might even alert predators to the youngsters' presence.

Noisy play: risk or signal?

Squirrel monkeys and their close relatives the tamarins are among the few primates who engage in noisy play (Biben & Symmes 1986; Goedeking & Immelmann 1986). The vocalizations made during play indicate the motivation or willingness to continue a bout of play, information which *could* be used by the partner, although neither study found actual evidence that it *was* used. Most play vocalizations are given by the individual in the defensive position in a play fight, the one on the bottom, and they are an honest indicator of willingness to continue.

Squirrel monkey play peeps are loud, only about 5 dB softer than their 'lost' vocalization, suggesting that they might be intended for other ears, for instance, those of nearby adults. And if the adults are informed by this broadcast signal, then it is likely that any predators in the vicinity will also be informed. These signals are loud and have an abrupt frequency sweep, features that make them easily localizable and put the whole troop at risk. Risk of predation is just one of many hazards to which play exposes a youngster.

Squirrel monkey parents appear to do little or no monitoring of the safety of their youngsters' play environment, but they do become more vigilant for predators when youngsters play (Biben et al. 1989). Play vocalizations alone are sufficient to cause the increase in vigilance, apart from any visual cues of play. This kind of indirect evidence of the risks of play is about as good as we have at the present time but it is an important variable on the *cost* side in the often discussed cost/benefit ratio of play, energetic cost being the most often cited.

Energy expenditure

Saimiri is among the few genera where the deleterious effect of energy shortage on the occurrence of play has been demonstrated in both the field and the laboratory (Baldwin & Baldwin 1973, 1974, 1976). Under benign conditions, squirrel monkeys *do* expend a lot of time and what looks like a lot of energy on play, but then these are very active Cebids. Of more relevance than risk incurred by including play in the repertoire are risks (other than physical) incurred during play itself .

Social risks

In children, role reversal is not always observed, and repeated participation in one-sided play may have negative social or psychological consequences for some participants (Sutton-Smith & Kelly-Byrne 1984). The possibility of animals' play having differential advantages for interactants has been proposed in theory (Fagen 1981, 1984) but little evidence has been put forth in support.

Males' playfighting may benefit some individuals at the expense of others. Clearly, we see young squirrel monkeys striving to be dominant during play and we can document that some are more successful in this than others, over a fairly long period. It is equally clear that dominance during play derives from dominance relationships outside of play, and that individuals do not improve their nonplay dominance status through play, at least in the short term. This might suggest that play fighting merely reinforces a relatively immutable dominance relationship between two squirrel monkey youngsters, as seems to be the case in preadolescent boys (Neill 1976).

Play fighting as training for serious fighting

The similarities between play fighting and serious fighting are a keystone of most play-as-practice theories, for here are activities sharing both behavior patterns and differential gender participation. Fighting requires skill, and has important immediate (social) and long term (fitness) outcomes. Play fighting is the only juvenile activity that resembles fighting, and it is performed with great frequency. Clearly, practice should be beneficial, if not absolutely necessary, and play is the only candidate on the horizon. So play must be practice for fighting, right?

Too literal a reading of the role of play fights as practice for fighting is perhaps the result of too casual an interpretation of the design feature paradigm of play (Smith 1982). As discussed by Caro in his 1988 review of the state of research on the adaptive significance of play, most of our information on the benefits or adaptive significance of play has come from observational studies of the form or design of youngsters' play, most of which conclude that play is in some way practice or preparation for the adult behaviors they resemble. Play fighting in squirrel monkeys, like play fighting in many other primates and the rough and tumble activity of children, shares patterns closely with serious fighting and is performed more often and more roughly by males, who do most of the serious fighting.

Nonetheless, as has been reiterated many times in the literature, play may be distinguished from serious fighting (Aldis 1975; Humphreys & Smith 1987; Pellis & Pellis, Chapter 6) with few exceptions (Hole & Einon 1984). Juvenile and adult animals and all but the most novice human observers can make the distinction. Only adult humans observing juvenile humans seem uncertain, leading to discouragement of rough and tumble on playgrounds in Western societies (Sutton-Smith 1993). While the existence of the differences has been documented extensively, their significance has remained, I believe, unappreciated.

Certainly, *play* fighting would seem to provide opportunities to learn about *real* fighting. But how good is this practice really? Evidence that opportunity to *play* at an activity is essential or even helpful in achieving adult proficiency is lacking (Chalmers & Locke-Haydon 1984; Caro 1995). Descriptions of actual fights in adult squirrel monkeys are hard to come by, but one thing is certain: a major characteristic is the inflicting of bites. Yet we never saw biting, inhibited or otherwise, or wounding in play, making play look like a less-than-perfect model for at least this important aspect of fighting. Other motor skills for combat do appear to be present in useable form but playful changes in their sequencing may render their performance of dubious value for practicing how to fight. It is now 30 years since Loizos (1966) pointed out that at least some of the characteristics that make play fighting playful (different sequencing of behaviors, inhibition of wounding) are counterproductive to learning to fight effectively. More recently, Stamps (1995) reiterated that only the execution of actual movements and sequences of movements are effective practice. Sutton-Smith (1995) includes physical forms of play in his conclusion that the evidence is lacking for the transfer of skills from play to any specific serious activity. At least with respect to motor patterns,

Sutton-Smith (1995, p. 281) may not be far wrong in suggesting that '... what one learns from play is how to play better.' So, if the motor patterns and the sequencing of motor patterns is less than ideal, and maybe even counterproductive to learning to fight, then what is going on with this activity that looks so much like fighting and yet is an apparently flawed imitation of it?

Another major point of departure is role reversal: in a real fight, animals give their opponent no quarter. One function of role reversal is to keep play bouts going, but intentionally losing is not what happens in a real fight. How then can role reversal be practice for real fighting? To answer this we need only look at the reality of growing up in a polygynous society, where males compete for females, food, and space. Conventionally, male–male competitiveness is viewed as fighting, displaying, intimidating, and alliance formation. The reality may not be so glorious. In most primate societies, males work up to the top, during years in which their competition is likely to be older, bigger, stronger, and more experienced than them. It would seem to be beneficial for any young male monkey, including those dominant among peers, to be able to judge the competition and to learn where to pick fights. And even more beneficial, should he find himself engaged in a mismatch, would be to not panic, but to assume the subordinate role and make the best of a bad situation. While the motor aspects of a display of subordinance are not difficult to perform, this is, after all, a position in which one is maximally vulnerable to injury, should the dominant animal choose to inflict it. 'For, those that fly, may fight again, Which he can never do that's slain', credited to Samuel Butler, is one of several variations on a theme that reoccurs many times throughout Western literature, but which has received much less attention in the ethological literature. Monkeys cannot be expected to be philosophical about losing, but it is important for a young male to maintain his confidence even as he displays his subordinance in his body posture.

When a juvenile male gives up, in play, opportunities to be dominant, he may gain experience in both the motor and cognitive aspects of being subordinate - experience that may later save his hide for another day. And because dominance roles during and outside of play are separate, he can do so without abdicating his normally dominant role.

Perhaps these lessons are learned more effectively, and with less association with fear, in the low-stress atmosphere of play, rather than in more stressful 'real' dominance or agonistic encounters. Combat, sexual behavior, and predation involve risky close-quarters behaviors whose stress-inducing properties may be reduced through similar encounters in play (Martin & Caro 1985; Pellis & Pellis 1996). Thus, play fights may be less accurately considered just as practice for *fighting* and more accurately viewed as generally applicable practice for social encounters, particularly intense close quarters ones, including combat related aspects such as sizing up an opponent, and bravery.

Based on these observations of wrestling play in young male squirrel monkeys, I speculated (Biben 1989) that without playful experience in the dominant role, young monkeys might grow up to become sissies and without play in the subordinate role, they might become bullies. Controlled longitudinal studies to test these predictions are difficult to achieve, but data from studies of rat play reveal some intriguing associations between early play experience and later responses to insult that might be termed either thin-skinned, or sissy.

In an experiment designed to differentiate the effects of play deprivation from the more general effects of social isolation, Potegal & Einon (1989) compared the aggressive responses of rats whose only social experience was through play, rats having neither social nor play experience, and a control group of rats raised with social companions with whom they also played. Periods of social isolation including, of course, play fighting, are associated in rats with either hyper- or hypo- defensiveness (measured as defensive aggressive response to painful stimulation) depending on the length and timing of the isolation. Whereas isolated rats were socially incompetent in potentially aggressive encounters, taking offense where none was intended and failing to respond to real threats, a daily hour of play experience was sufficient to prevent or reverse these effects.

Studies of the playground rough and tumble of adolescent boys support an association between antisocial play strategies and social competence. Rejected (unpopular) boys, who lack the skills to engage in the cooperative games that build social support and status in adolescence, are much more likely than peers to express dominance and aggression in their play (Olweus 1979; Humphreys & Smith 1987; Pelligrini 1995b). More than others, rejected boys use rough and tumble to bully and victimize vulnerable lower-ranking children. Of course, antisocial play habits are only one aspect of more general social incompetence in some children, and not necessarily a cause of it (Sutton-Smith 1995).

A bully's overresponse to a presumed insult and the hypersensitive self-protection of a sissy are rarely successful long-term strategies in a complex society (Olweus 1991, 1992). Even before children are old enough to rough and tumble, they may be learning lessons about mock and real attacks through tickling play. Infants initially respond to tickling as if they were under attack, stiffening, frowning and crying. After repeated tickling games, the child learns that these are only mock attacks and begins to enjoy them (Chance 1979).

The social competence benefits of play fighting may include short term as well as long term advantages. Research on elementary school aged children suggests that participation in rough and tumble play, which is almost exclusively the domain of boys in any culture, confers social competence advantages such as popularity and the opportunity to engage in affiliative behavior and games (Pelligrini 1988, 1995a). Young boys avoid play partners whose dominance rank differs greatly from their own and, as with squirrel monkeys, young children do not improve their dominance situation in a group through their performance in rough and tumble play. As boys get older, rough and tumble is used to bully and display dominance. It declines in frequency, and becomes negatively associated with popularity (Pelligrini 1995b).

So, we return to the question of what, if anything, play fighting might have to do with real fighting? A lot, as it turns out:

1. Behavioral flexibility

Because play has more variability than serious activity, it has long been assumed that play is about flexibility (Bruner 1972). While there is surprisingly little evidence to support the flexibility hypothesis, Pelligrini (1995a) has recently demonstrated a correlation between boys' ability to switch activities during rough and tumble play, and their later scores on a test of social problem-solving ability. Ability to use and switch among alternative sequences may be as valuable as getting a lot of practice at the most effective sequences.

2. Becoming expert at reading intentions of others

Bekoff & Allen (Chapter 5) have speculated that, despite the presumed deficiency of nonhuman animals in perceiving the mental states of others, play may promote learning about the intentions of others. The ability to observe, understand, and act upon the intentions of an opponent during a fight is essential. And it may be even more important to judge the intentions of a potential opponent: i.e., is he serious or bluffing, how motivated is he, is this going to be play or is it for real? Engaging in wrestling play provides both kinesthetic and cognitive cues to the partner's (and one's own) ability or likelihood to move from one position to another.

3. Reducing the stress of close bodily contact

Play fighting is by definition a contact activity, and close contact with strangers or those whose intentions are unclear is of course another stressor. Hence the concept of 'personal space' and the necessity of mating rituals, greeting ceremonies, etc., to breech it. Play provides plenty of exposure to close quarters activity in an atmosphere of fun. The association with fun, as well as the repeated participation in the activity, may reduce animals' distress at being in potentially injurious contact (whether agonistic or sexual), and give them confidence.

4. Experience in both the dominant and subordinate roles

The act of pinning, with one animal (in the dominant role) forcefully holding down the other (in the subordinate role) is perhaps the closest of all close quarters contact in play fighting.

Being subordinate can be stressful (Sapolsky 1993) and actually being forced by a dominant animal to assume a subordinate posture is probably very stressful. Because all youngsters are subordinate to all adults in a squirrel monkey troop, it is likely that all have exposure to this role in a 'real' context. Playing at being subordinate may have both present and long term benefits for an animal by making this experience less stressful when necessitated during a fight or other agonistic encounter. The additional benefits of learning how to defend against attack, discussed by Pellis & Pellis (Chapter 6), are most readily accessed by the interactant in the defensive or subordinate position.

Expectations of dominance are different for squirrel monkey males and females, as in most primates. The philopatric females appear to maintain fairly stable dominance relationships among themselves, which are evident even among very young females. Typically, females are subordinate to all adult males. Males experience many changes in their dominance relationships as they mature, emigrate (probably forcibly) from their natal troop, become part of a bachelor group, and, finally, establish themselves in a new troop. Of course, not all males will achieve this success. To accrue whatever benefits dominant status may confer, a male's adeptness at negotiating these moves and handling the stress associated with them is vitally important. And because youngsters, even males, hold subordinate positions in the troop, experience at being dominant (or forcing someone else to be subordinate) is hard to find outside of play.

5. Bravery

Discussing the mechanism by which play fighting experience might avoid the hypersensitivity to painful stimulation observed in rats raised in social isolation, Potegal & Einon (1989) speculated that participation in rough play might train animals to tolerate painful stimulation during social interaction. They suggested that the adaptive significance of such tolerance might be to reduce the number of agonistic encounters that an animal is drawn into, presumably by making him more tolerant of insult. For real battles, tolerance to pain should make him more persistent or 'brave'.

Biting is not a frequent activity in squirrel monkey play fights, and is actually not common even in rats' play (Pellis & Pellis 1987), which does not, of course, diminish its importance as a stimulus. Nonetheless, biting is not the only stressful activity an animal may be habituated to in play. The close-quarters contact experience discussed above is also likely to be important in giving an animal courage.

In the conventional view of play fighting as practice for fighting, it has generally been implied that what is being learned is how to be aggressive, but there is little evidence to support this rather hasty conclusion. Instead, a more careful examination of play fighting behavior in rodents (Meaney & Stewart 1981; Pellis & Pellis 1987, and Chapter 6) suggests that the maneuvers used in play fights are more relevant to defense than to attack. Augmenting this view, I would here suggest that what squirrel monkeys, particularly young males, are learning in play fighting are indeed lessons relevant to fighting. These are not instruction in combat maneuvers nor are they even primarily about aggression. Instead, through their play strategies, and partner choices squirrel monkeys are learning cognitive lessons relevant to social competence and self preservation, lessons of such importance that the opportunity to play can compel youngsters to disregard preferences and embrace any potential partner providing an opportunity to play, even if it means adjusting the game or play style.

References

Aldis, O. 1975. *Play Fighting*. New York: Academic Press.

Altmann, S. A. 1962. Social behavior of anthropoid primates: Analysis of recent concepts. In *Roots of Behavior* (ed. E. L. Bliss), pp. 277–85. New York: Harper & Brothers.

Baldwin, J. D. 1969. The ontogeny of social behaviour of squirrel monkeys (*Saimiri sciureus*) in a seminatural environment. *Folia Primatol.*, **11**, 35–79.

Baldwin, J. D. 1971. The social organization of a semifree-ranging troop of squirrel monkeys (*Saimiri sciureus*). *Folia Primatologica*, **14**, 23–50.

Baldwin, J. D. & Baldwin, J. I. 1973. The role of play in social organization: comparative observations on squirrel monkeys (*Saimiri*). *Primates*, **14**, 369–81.

Baldwin, J. D. & Baldwin, J. I. 1974. Exploration and social play in squirrel monkeys (*Saimiri*). *American Zoologist*, **14**, 303–14.

Baldwin, J. D. & Baldwin, J. I. 1976. Effects of food ecology on social play: A laboratory simulation. *Zeitschrift fur Tierpsychologie*, **40**, 1–14.

Bekoff, M. 1976. The social deprivation paradigm: who's being deprived of what? *Developmental Psychobiology*, **9**, 497–8.

Bekoff, M. 1977. Social communication in canids: evidence for the evolution of a stereotyped mammalian display. *Science*, **197**, 1097–9.

Bekoff, M. 1978. Social play: structure, function, and the evolution of a cooperative social behavior. In: *The Development of Behavior: Comparative and Evolutionary Aspects* (eds. G. Burghardt and M. Bekoff), pp. 367–83. New York: Garland.

Bekoff, M. 1995. Play signals as punctuation: The structure of social play in canids. *Behaviour*, **132**, 419–29.

Biben, M. 1986. Individual- and sex-related strategies of wrestling play in captive squirrel monkeys. *Ethology*, **71**, 229–41.

Biben, M. 1989. Effects of social environment on play in squirrel monkeys: resolving Harlequin's Dilemma. *Ethology*, **81**, 72–82.

Biben, M. & Champoux, M. 1996. Evidence for an inverse relationship between stress and play. Presented at the annual meeting of the Animal Behavior Society, Flagstaff, Arizona.

Biben, M. & Suomi, S. 1993. Lessons from primate play. In: *Parent-Child Play: Descriptions and Implications*. (ed. K. MacDonald), pp.185–96. Albany, NY: State University of New York Press.

Biben, M. & Symmes, D. 1986. Play vocalizations of squirrel monkeys (*Saimiri sciureus*). *Folia primatol.*, **46**, 173–82.

Biben, M., Symmes, D. & Bernhards, D. 1989. Vigilance during play in squirrel monkeys. *Am. J. Primatol.*, **17**, 41–9.

Boinski, S. 1987. Birth synchrony in squirrel monkeys (*Saimiri oerstedi*). *Behav. Ecol. Sociobiol.*, **37**, 393–400.

Boulton, M. 1991. Partner preferences in middle school children's playful fighting and chasing: A test of some competing functional hypotheses. *Ethology and Sociobiology*, **12**, 177–93.

Bruner, J. 1972. The nature and uses of immaturity. *American Psychologist*, **27**, 687–708.

Byers, J. 1980. Play partner preferences in Siberian ibex, *Capra ibex siberica*. *Zeitschrift fur Tierpsychologie*, **53**, 23–40.

Caro, T. M. 1988. Adaptive significance of play: are we getting closer? *Trends in Ecology and Evolution*, **3**, 50–3.

Caro, T. M. 1995. Short term costs and correlates of play in cheetahs. *Animal Behaviour*, **49**, 333–46.

Chalmers, N. & Locke-Haydon, J. 1984. Correlations among measures of playfulness and skillfulness in captive common marmosets (*Callithrix jacchus jacchus*). *Developmental Psychobiology*, **17**, 191–208.

Chance, P. 1979. *Learning through Play*. New York: Gardner Press.

Cheney, D. 1978. The play partners of immature baboons. *Animal Behaviour*, **26**, 1038–50.

DuMond, F. 1968. The squirrel monkey in a seminatural environment. In: *The Squirrel Monkey* (eds. L. Rosenblum & R. Cooper), pp. 87–145. New York: Academic Press.

Einon, D. F., Morgan, M. J. & Kibbler, C. C. 1978. Brief periods of socialization and later behavior in the rat. *Developmental Psychobiology*, **11**, 213–25.

Erwin, P. 1993. *Friendship and Peer Relations in Children*. Chichester, England: John Wiley & Sons, Ltd.

Fagen, R. 1981. *Animal Play Behavior*. New York: Oxford University Press.

Fagen, R. 1984. Play and behavioural flexibility. In: *Play in Animals and Humans*. (ed. P. K. Smith), pp. 159–73. Oxford: Basil Blackwell.

Fagen, R. 1992. Play, fun, and communication of well-being. *Play & Culture*, **5**, 40–58.

Fagen, R. 1995. Animal play, games of angels, biology and Brian. In: *The Future of Play Theory: A Multidisciplinary Inquiry into the Contributions of Brian Sutton-Smith*. (ed. A. D. Pelligrini). Albany, NY: State University of New York Press.

Goedeking, P. & Immelmann, K. 1986. Vocal cues in cotton-top tamarin play vocalizations. *Ethology*, **73**, 219–24.

Goodall, J. 1986. *The Chimpanzees of Gombe: Patterns of Behavior*. Cambridge, MA: Belknap Press.

Goodall, J. 1990. *Through a Window: My Thirty Years with the Chimpanzees of Gombe*. Boston: Houghton Mifflin.

Hole, G. & Einon, D. 1984. Play in rodents. In: *Play in Animals and Humans*. (ed. P. K. Smith), pp. 85–117. Oxford: Basil Blackwell.

Howes, C. 1988. Same- and cross-sex friends: Implications for interaction and social skills. *Early Childhood Research Quarterly*, **3**, 21–37.

Humphreys, A. P. & Smith, P. K. 1984. Rough-and-tumble in preschool and playground. In: *Play in Animals and Humans*. (ed. P. K. Smith), pp. 241–66. Oxford: Basil Blackwell.

Humphreys, A. P. & Smith, P. K. 1987. Rough and tumble, friendship, and dominance in schoolchildren: Evidence for continuity and change with age. *Child Development*, **58**, 201–12.

Jolly, A. 1985. *The Evolution of Primate Behavior*. New York: Macmillan.

Loizos, C. 1966. Play in mammals. *Symp. Zool. Soc. Lond.*, **18**, 1–9.

Martin, P. & Caro, T. M. 1985. On the functions of play and its role in behavioral development. *Advances in the Study of Behavior*, **15**, 59–103.

Meaney, M. J. & Stewart, J. 1981. A descriptive study of social development in the rat (*Rattus norvegicus*). *Animal Behaviour*, **29**, 34–45.

Neill, S. 1976. Aggressive and non-aggressive fighting in 12 to 13 year old preadolescent boys. *Journal of Child Psychology and Psychiatry*, **17**, 213–20.

Olweus, D. 1979. Stability and aggressive reaction patterns in males: A review. *Psychological Bulletin*, **86**, 852–75.

Olweus, D. 1991. Bully/victim problems among schoolchildren: Basic facts and effects of a school based intervention program. In: *The Develoment and Treatment of Childhood Aggression*. (ed. D. Pepler & K. Rubin), pp. 411–48. Hillsdale, NJ: Lawrence Erlbaum Associates, Inc.

Olweus, D. 1992. Bullying among schoolchildren: Intervention and prevention. In: *Aggression and Violence Throughout the Life Span*. (eds. R. D. Peters, J. McMahon & V. Quinsey), pp. 100–25. Newbury Park, CA: Sage Publications.

Panksepp, J,, Herman, B., Vilberg, T., Bishop, P. & DeEskinazi, F. 1978. Endogenous opioids and social behavior. *Neuroscience and Biobehavioral Reviews,* **4**, 473–87.

Pelligrini, A. D. 1988. Rough-and-tumble play and social competence. *Developmental Psychology*, **24**, 802–6.

Pelligrini, A. D. 1989. What is a category? The case of rough and tumble play. *Ethology and Sociobiology*, **10**, 221–341.

Pelligrini, A. D. 1995a. Boys' rough-and-tumble play and social competence: Contemporaneous and longitudinal relations. In: *The Future of Play Theory: A Multidisciplinary Inquiry into the Contributions of Brian Sutton-Smith.* (ed. A. D. Pelligrini), pp. 107–26. Albany, NY: State University of New York Press.

Pelligrini, A. D. 1995b. Adolescent boys' rough-and-tumble play. In: School Recess and Playground Behavior: Educational and Developmental Roles. (ed. A. D. Pelligrini), pp. 161–79. Albany, NY: State University of New York Press.

Pellis, S. M. 1988. Agonistic versus amicable targets of attack and defense: consequences for the origin, function, and descriptive classification of play-fighting. *Aggressive Behavior*, **14**, 85–104.

Pellis, S. M. & Pellis, V. C. 1987. Play-fighting differs from serious fighting in both target of attack and tactics of fighting in the laboratory rat *Rattus norvegicus. Aggressive Behavior*, **13**, 227–42.

Pellis, S. M. & Pellis, V. C. 1996. On knowing it's only play: the role of play signals in play fighting. *Aggression and Violent Behavior*, **1**, 249–68.

Potegal, M. & Einon, D. 1989. Aggressive behaviors in adult rats deprived of playfighting experience as juveniles. *Developmental Psychobiology*, **22**, 159–72.

Sapolsky, R. M. 1993. Endocrinology alfresco: Psychoendocrine studies of wild baboons. *Recent Progress in Hormone Research*, **48**, 437–68.

Smith, P. K. 1982. Does play matter? Functional and evolutionary aspects of animal and human play. *Brain and Behavioral Sciences*, **5**, 139–55.

Stamps, J. 1995. Motor learning and the value of familiar space. *American Naturalist*, **146**, 41–58.

Sutton-Smith, B. 1993. Dilemmas in adult play with children. In: *Parent-Child Play: Descriptions and Implications*. (ed. K. MacDonald), pp. 15–40. Albany, NY: State University of New York Press.

Sutton-Smith, B. 1995. Conclusion: the persuasive rhetorics of play. In: *The Future of Play Theory: A Multidisciplinary Inquiry into the Contributions of Brian Sutton-Smith*. (ed. A. D. Pelligrini), pp. 275– 95. Albany, NY: State University of New York Press.

Sutton-Smith, B. & Kelly-Byrne, D. (1984). The idealization of play. In: *Play in Animals and Humans* (ed. P. K. Smith). Oxford: Basil Blackwell.

Symons, D. 1978. *Play and Aggression: A Study of Rhesus Monkeys*. New York: Columbia University Press.

Thor, D. H. & Holloway, W. R., Jr. 1984. Social play in rats: A decade of methodological and experimental research. *Neuroscience and Biobehavioral Reviews*, **8**, 455–464.

Walters, J. 1987. Transition to adulthood. In: *Primate Societies* (eds. B. Smuts, D. Cheney, R. Seyfarth, R. Wrangham & T. Struhsaker). Chicago: University of Chicago Press.

Whiting, B. & Edwards, C. P. 1973. A cross-cultural analysis of sex differences in the behavior of children aged three through eleven. *Journal of Social Psychology*, **91**, 171–88.

9

Self assessment in juvenile play

KATERINA V. THOMPSON
Biological Sciences Program, University of Maryland, College Park, MD 20742 USA

Introduction

Many young mammals play and we still do not really understand why. The play literature abounds with hypotheses and speculation (Fagen 1981, Smith 1982, Baldwin 1986, Martin & Caro 1985), but quantitative support for most functional hypotheses is in frustratingly short supply. There is a wealth of descriptive data, but most of it deals with very basic and superficial aspects of play, such as sex differences in play frequency and the resemblance between play and serious fighting on a very gross level. Studies of play rarely have a strong theoretical underpinning. Undoubtedly, part of the problem results from the lack of precisely worded, falsifiable hypotheses (Caro 1988, Byers & Walker 1995). Compounding this is a tendency for certain ideas to become generally accepted, despite the lack of strong supporting data (e.g., that play is rehearsal for adult behaviors or serves to strengthen social bonds between playmates).

Some of the strongest recent play research addresses topics that are tangential to the function of play, for example the metabolic and fitness costs of play (Caro 1988, 1995; Miller & Byers 1991; Siviy & Atrens 1992) and the relationship between play and perturbations in the social environment such as decreased maternal responsiveness and the onset of weaning (Bateson et al. 1990, Smith 1991, Moore & Power 1992, Terranova & Laviola 1995). There is a need for carefully formulated hypotheses that generate testable predictions and detailed, quantitative data that are directly relevant to tests of these predictions.

In this paper, I will briefly consider several aspects of play that seem inadequately explained by current theory. These include (1) the possible implications of the brief, repetitive nature of play behaviors, (2) whether or not play is a unitary category, (3) the ambiguous relationship between play and aggression, and (4) the question of whether play is competitive. I will then suggest an alternative interpretation of play: that play is a mechanism by which a developing individual can assess its capabilities.

Challenges for current theory

Why are play bouts so brief?

Play can be characterized by very brief, discrete elements that tend to be repeated during individual play bouts (Loizos 1966). Individual play behavior patterns and sometimes bouts themselves are often only a few seconds in duration. For example, a quick survey of the literature shows mean durations for play fighting in the range of 5 to 20 seconds for a wide variety of species (baboons, *Papio anubis*: Owens 1975a; mountain goats, *Oreamnos americanus*: Dane 1977; rhesus macaques, *Macaca mulatta*: Symons 1978; punarés, *Thrichomys apereoides*: Thompson 1985; human children: Fry 1987; laboratory rats, *Rattus norvegicus*: Hole 1988; red-necked wallabies, *Macropus rufogriseus banksianus*: Watson & Croft 1993).

This brevity runs counter to the predictions of functional hypotheses that suggest that play is responsible for optimizing skeleto-muscular development and increasing endurance (Fagen 1976). Optimal development of strength and endurance is achieved when muscles are exercised to near exhaustion, therefore play bout durations should be longer than expected if play bout terminations were determined by random events. The distribution of play bout durations differs from that predicted by chance in the locomotor play of ponies, *Equus caballus* (Fagen & George 1977) and domestic cats, *Felis catus* (Fagen 1981), but not in the play fighting of California ground squirrels, *Spermophilus beecheyi* (McDonald 1977), laboratory rats (Hole 1988) or rhesus macaques (Fagen 1981). Perhaps the critical characteristic of play behavior patterns is not their duration so much as their repetitive nature.

Is play a unitary category?

Play has traditionally been classified into three basic categories: (1) object play, which involves the manipulation of inanimate things; (2) locomotor play, which consists of activities such as running and rotational body movements; and (3) social play, which involves two or more individuals that respond to each other's actions (Fagen 1981). There has been some movement in the literature to try to consider these types of play separately because they develop at different times in ontogeny (Gomendio 1988) and appear to differ in their motivational basis (Pellis 1991). Furthermore, some of the explanations offered for play pertain only to social play (e.g., social bonding, Carpenter 1934).

This approach at first seems very attractive because it subdivides an unwieldy research topic into smaller, seemingly more manageable units. However, some descriptions of play suggest that object, locomotor and social variants form a single natural category. First, different types of play tend to be interspersed within the same play bout. Locomotor play in sable antelope, *Hippotragus niger*, calves that begins as a solitary run by a single calf may segue into a complex chase involving several infants (pers. obs.). Individual play sequences in mountain goat kids span a continuum from solitary acrobatic maneuvers ('whirling' and jumping), to non-contact group activities ('King of the Castle') to contact social play behaviors such as butting (Dane 1977). Solitary and social elements are similarly combined in the play of human children (Fry 1987), infant harbor seals, *Phoca vitulina* (Renouf & Lawson 1986), Columbian ground squirrels, *Spermophilus columbianus* (Steiner 1971) and laboratory rats (Pellis & Pellis 1983).

Second, there are many examples of play that defy simple classification. Biben (1982, 1983) has reported that object play in infant bush dogs, *Speothos venaticus*, has a strong social component. Seventy-seven percent of object play bouts consist of groups of pups jointly carrying around a single stick or rock. In ungulates, group locomotor play also appears to be social in nature because infants respond to each other's actions by chasing and dodging, making this form of play as socially interactive as play fighting. In the Zapotec-speaking people of Mexico, children combine object and locomotor play by tying their feet together with ribbon during locomotor play bouts (Fry 1987). Object, locomotor and social elements are all present in a children's bicycle race. Rather than being different types of behaviors that have been artificially lumped together, object, locomotor and social play evidently share some elusive characteristics that make them recognizable by even casual observers as being different forms of the same phenomenon. Hypotheses that deal with only one of these forms are unlikely to provide the foundation for understanding the diversity of play. Any truly satisfying explanation for play needs to account for all three of these forms.

The relationship between play and aggression

The resemblance between play and aggression has long been cited as support for the idea that play serves as rehearsal for behaviors critical to reproduction and survival in the adult (Groos 1898, Symons 1978, Fagen 1981, Smith 1982). Upon close examination, however, play appears to differ from aggression in ways that are fundamentally at odds with predictions derived from this 'practice' hypothesis. First, the specific targets of attack and defense in play and serious fighting may differ (Pellis 1988, 1993). Second, the most challenging elements of serious fighting are sometimes absent from play fighting, when practice would seem crucial to their successful performance (Pellis & Pellis, Chapter 6). Finally, in many species infants have the capability of performing aggressive behaviors in their adult forms prior to or coincident with the first emergence of social play, indicating that playful practice is not necessary for their expression (polecats, *Mustela putorius*: Poole 1966; canids: Bekoff 1974; baboons: Owens 1975b; punarés: Thompson 1985; spotted hyenas, *Crocuta crocuta*: Drea et al. 1996). At any point in an individual's life history, play and aggression are structurally distinct and reliably distinguishable (Fig. 9.1).

Is play competitive?

Winners and losers of individual play bouts are often discernible

In many species, it is possible to designate one individual in a playing dyad as the 'winner' based upon postural differences and its degree of control over subsequent events. For species that wrestle, the winner is

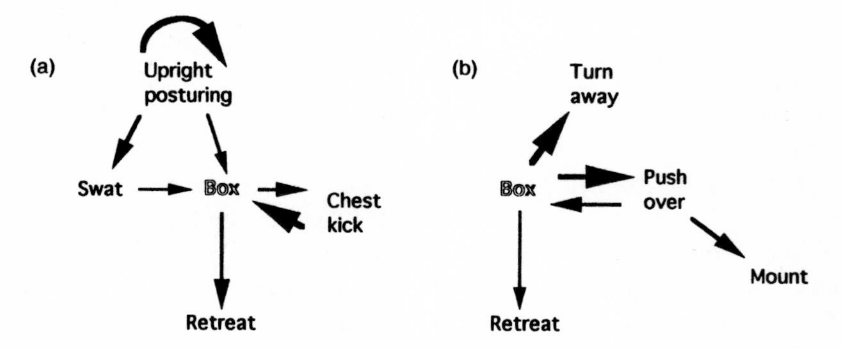

Fig. 9.1. Sequences of behavior during serious fighting (a) and play (b) in infant punarés. The width of arrows is proportional to the likelihood of a transition between two behaviors.

usually the individual on top, and the behavior is often referred to as pinning (Columbian ground squirrels: Steiner 1971; laboratory rats: Panksepp 1981; canids: Biben 1983; squirrel monkeys, *Saimiri sciureus*: Biben 1986; yellow-bellied marmots, *Marmota flaviventris*: Jamieson & Armitage 1987; human children: Fry 1987). In other species, the winner can be designated as the individual that succeeds in causing its partner to lose balance and fall (punarés, Fig. 9.2; wolves, *Canis lupus*: Havkin & Fentress 1985). These positional differences among play partners are not random occurrences, but rather are the outcome of fairly sophisticated maneuvers that are consistent with a strategy of attempting to gain physical advantage over the partner (Havkin & Fentress 1985).

In punarés, the winner of a play bout is much more likely to mount the loser than vice versa (Fig. 9.3). This may simply be due to the postural advantages of remaining upright when the other individual is knocked to the substrate. Another correlate of winning is a behavior once referred to as a 'frisky hop' (Wilson & Kleiman 1974). It consists of jumping directly upwards from a quadrupedal or bipedal stance while simultaneously shaking the head and is often repeated in quick succession. Although sometimes seen in the context of social play, it is not a behavior that involves interaction with the play partner. Play bout winners are three times more likely to exhibit frisky hops than are losers. This behavior seems to occur when the animals are highly excited, during both solitary locomotor play and social play, and may be an indication that asymmetries in the outcome of social play bouts are recognizable by the participants.

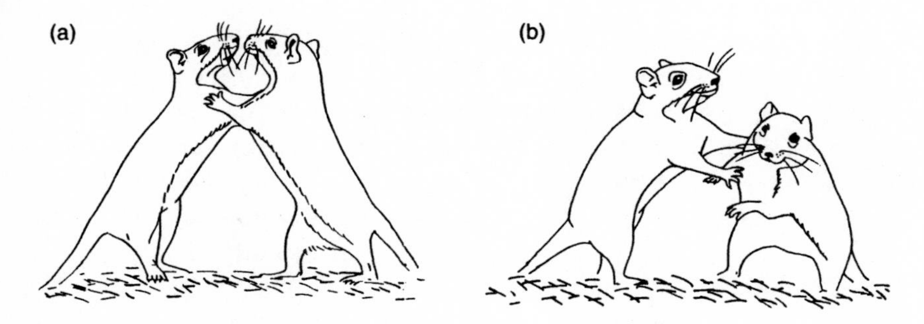

Fig. 9.2. Typical social play patterns in infant punarés. (a) Boxing is the predominant play behavior pattern, and consists of two individuals in a bipedal stance pushing against each other's shoulders with the forepaws. (b) Push over occurs when one individual pushes the other off balance and onto the substrate.

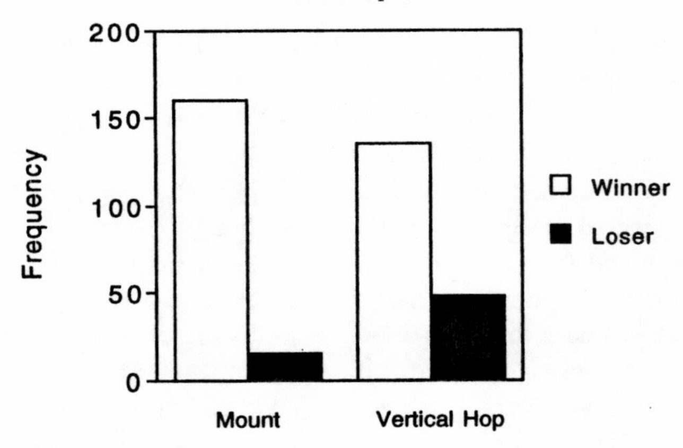

Fig. 9.3. Frequency of Mounting and Frisky (or Vertical) Hopping by winners and losers of play bouts. Losers were defined as individuals who terminated play interactions by Turning away or Retreating, or who were Pushed over onto the substrate by their partners.

Self handicapping

Self handicapping is said to occur when the stronger, bigger or more skilled of two mis-matched play partners adjusts its play intensity to match that of the other individual (Aldis 1975, Symons 1978, Watson & Croft 1996; see also Bekoff & Allen, Chapter 5, and Biben, Chapter 8). Evidence for the existence of self handicapping is very weak and rests largely on logical arguments, rather than quantitative evidence. One piece of evidence used to support the existence of self handicapping is the infrequency of predictable, asymmetrical win–loss relationships during play (Aldis 1975). In a given pair of individuals, dominance relationships may be very well defined outside of play but very ambiguous during play (Owens 1975b; Biben 1986, 1989; Watson 1993). That is, one individual consistently 'wins' dominance interactions, but during play that individual wins much less frequently. This could be interpreted as self handicapping: the dominant individual is allowing the subordinate to win occasionally.

The assumption implicit in this argument is that asymmetries in dominance interactions are a good indicator of the relative competitive abilities of the two individuals. This is unlikely, however, since dominance relationships are characterized by highly asymmetrical interactions that are maintained in large part because the subordinate avoids direct confrontation with the dominant individual (Rowell 1973, Bernstein 1981). A very strong, predictable dominance relationship may result from subtle differences in competitive abilities. An alternative interpretation of this same situation is that wins and losses in play are an accurate reflection of differences in competitive abilities. Dominance interactions are likely to be more highly asymmetrical because the subordinate habitually defers to the dominant.

This is not to imply that self handicapping never occurs. My subjective impression is that it does, especially between greatly mis-matched partners such as parent and offspring. The phenomenon of self handicapping needs more careful, quantitative analysis before its significance can be evaluated.

Watson & Croft (1996) make a strong case for the existence of self handicapping in adult male red-necked wallabies playing with much younger partners. They observed asymmetries in the degree of 'offensive' play maneuvers performed during play bouts, with the younger animals playing with greater intensity and the older individuals tending to stand flat-footed and reacting defensively. This suggests that it may be useful to compare the play of well-matched dyads to that of mis-matched dyads to look for asymmetries in intensity, restraint and use of specific motor patterns.

Role reversal

Another commonly made assumption is that role reversal (i.e., alternation of the dominant and subordinate positions) is an important objective of play, and that play partners self handicap in order to maintain some degree of role reversal (Aldis 1975; Biben 1986, 1989). While it is true that play bouts are often characterized by role reversal, this does not imply that reversals are evidence of self handicapping. Rather, it may simply be an indication that two play partners are closely matched (that is, similar in size, strength and skill). In many species, individuals express preferences for well-matched partners (sable antelope: Thompson 1996a; bighorn sheep, *Ovis canadensis*: Berger 1980; Siberian ibex, *Capra ibex siberica*: Byers 1980; rhesus macaques: Brueggemann 1978; common marmoset, *Callithrix jacchus*: Stevenson & Poole 1982; human children: Boulton 1991). In such situations, it is likely that each individual will find itself in the 'dominant' position at least part of the time, and self handicapping need not be invoked to explain the existence of role reversal.

If play partners were selected randomly however, then instances in which play partners were grossly mis-matched would be expected. In these situations, win–loss relationships would be expected to be highly asymmetrical unless one of the individuals self handicapped. For example, a common experimental paradigm in studies of social play is to house weanling rats individually and allow them brief, daily periods of social interaction. The primary activity during these contact periods is intense social play. If during these contact periods the rats are repeatedly paired with the same, randomly chosen partner, they develop highly asymmetrical win–loss play relationships that are stable over time (Panksepp 1981). Self handicapping is not readily apparent in this case.

Is winning important?

Juvenile baboons (Owens 1975b) and squirrel monkeys (Biben 1986) are more likely to initiate play with partners whom they can dominate. Biben (1986) observed that most play in squirrel monkeys occurred between individuals with similar probabilities of winning the interaction, and play between mis-matched partners was less frequent. When dominance relationships were highly polarized, play did not occur at all (Biben 1989).

Partners that are apparently mis-matched in some attributes may not be mis-matched in others, and this may allow stability in forms of play that rely on the more well-matched attributes. For example, two individuals of vastly different size may be mis-matched during wrestling but may be better matched in forms of play requiring agility. In squirrel monkeys, role reversal during wrestling on flat surfaces ('directional' wrestling) is less frequent in male–female dyads than in single-sex dyads (Biben 1986). Females infrequently participate in directional wrestling with males and instead engage in 'non-directional' wrestling whilst suspended from branches, where dominant and subordinate roles are less clearly defined. This implies that males are not self handicapping themselves during directional wrestling, and that females are responding to this situation by switching to a different form of play, in which the disparities between males and females are minimized. Biben (1986) has interpreted this as a strategy used by females to minimize their probability of losing play bouts. Experimentally formed dyads mis-matched for sex and/or age show a similar tendency to engage in non-directional wrestling (Biben 1989).

Infant punarés also seem to behave in such a way as to maximize their probability of winning play bouts. For every male dyad, the probability of winning was higher when the infant initiated the bout than when it was the recipient (Fig. 9.4). The initiator of a play bout may have a positional advantage that allows the recipient to be more easily pushed off balance. Initiating play is also correlated with achieving the dominant position in baboons (Owens 1975a) and laboratory rats (Hole 1988). In squirrel monkeys (Biben 1986) and laboratory rats (Hole 1988), there is little evidence of reciprocity and role reversal within individual play bouts. Once the dominant position is achieved, it is not readily relinquished.

One context that provides an opportunity to evaluate the importance of winning in play is the phenomenon of escalation of play bouts into aggression. This is very rare, since most play partners are related or at least members of the same social group, so there are virtually no quantitative data on the circumstances surrounding escalation. Escalation of play into agonism reliably occurs between unfamiliar, male infant punarés. I recorded five instances of escalation among 42 such staged encounters. In each case, the encounter followed a similar sequence of events (Fig. 9.5). Agonism appeared after win–loss relationships had been established in play and, in every instance, the first aggressive act was performed by the losing individual. Once agonism appeared, the infants never returned to play.

Social play appears to have a strongly competitive element. Winners and losers of play bouts are often easily distinguishable by human observers, and winners may differ from losers in the behaviors that follow play. Rather than encouraging role reversal, playing infants appear to

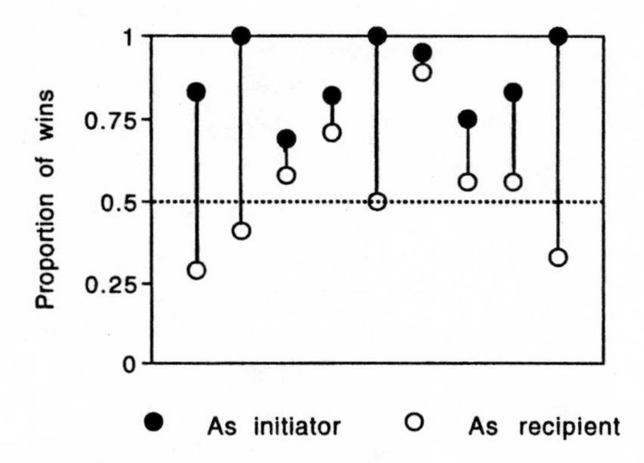

Fig. 9.4. Probability of winning a play bout when in the role of the initiator and recipient for nine sibling dyads of infant male punarés. Data are plotted from the perspective of the member of each dyad most likely to win.

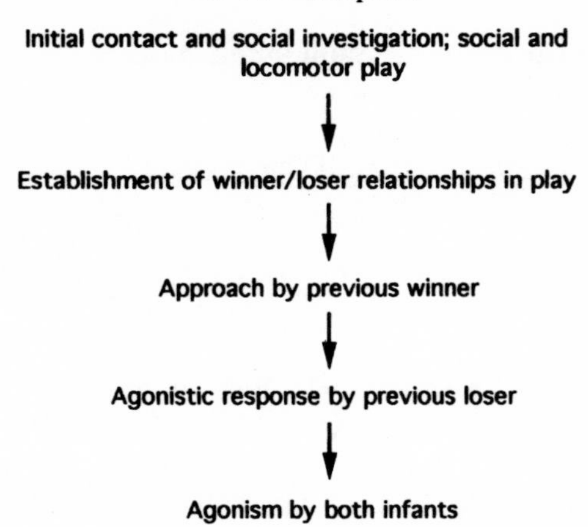

Fig. 9.5. Sequence of events during staged encounters between unfamiliar infant punarés.

use strategies to maximize their probability of winning play bouts. The importance of winning in play is inadequately explained by current functional hypotheses. I propose that the competitive nature of play is integral to its effectiveness.

Self assesment in play

If specific types and amounts of play experience are necessary for optimal development, then natural selection should favor the evolution of mechanisms for assuring that play (and thus development) proceeds optimally. I suggest that play is a mechanism of managing development. Play provides developing individuals with immediate feedback on their physical abilities, and this feedback can be used to regulate future activities. I will refer to this as self assessment.

Under this view, play consist of a series of self-selected activities (tests) that provide immediate feedback on an individual's competence at manipulative, locomotor or competitive social skills. This immediate feedback derives from an important property of the tests themselves: the presence of a dichotomous outcome. In the context of locomotor and object play, the outcome is success or failure; while in social play it is winning or losing. Two implications follow: locomotor play is inherently risky, because there is always a chance of failure, and social play is inherently competitive.

Tests are progressive in that, at any given point in time for a given individual, tests can be ranked on their probability of success (or, for social tests, the probability of winning). In ungulate locomotor play, for example, the simplest variant is running (Byers 1984). Locomotor play can become more complex by the addition of other balance challenging movements, such as simultaneously kicking up the heels or tossing the head. Success is completing the maneuver, while failure is manifested as a loss of balance, often to the point of falling (Dane 1977; Byers 1987; author's pers. obs.).

The outcome of tests is used as the basis for selecting subsequent tests such that play is 'managed' at some optimal outcome probability. For locomotor and object play, this means that repeated success at any given test results in the individual moving on to a test with a lower probability of success, while repeated failure results in switching to a test with a higher expectancy of success (Fig. 9.6). At some intermediate level, if the performance of a particular test is associated with some advantageous developmental effect, such as cerebellar synaptogenesis or muscle fiber-type differentiation (Byers & Walker 1995), the test will be repeated until the success ratio necessary for switching to another test is reached. Even if the test itself has no influence on physiological development, it is likely to be repeated. Each test can be viewed as a statistical 'sampling' of the individual's developmental state and as such is vulnerable to stochastic influences that might lead to an inaccurate assessment. Repeated testing

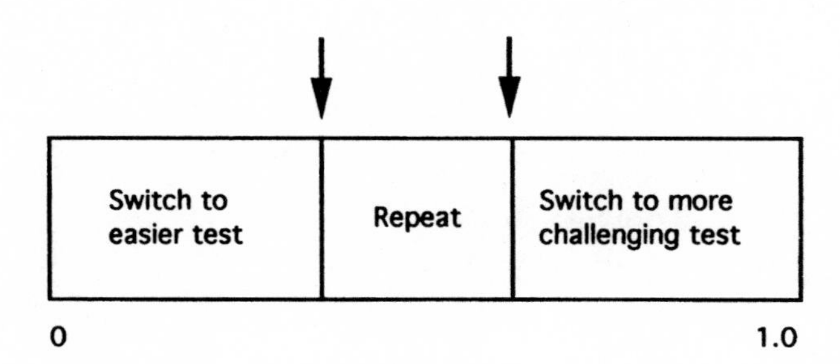

Fig. 9.6. Diagrammatic representation of the decision rules governing play under the self assessment hypothesis. Arrows indicate probability levels at which a playing individual switches from one test to another. See text for details.

assures that the decision to switch to a riskier test is based on an accurate assessment.

A brief example of a four-year old child playing with a basketball illustrates this principle. During one particular day, two variants were tried by the child: the standard one-handed dribble and attempting to bounce the ball under her leg. Success was defined as one complete sequence of the child hitting the ball, the ball bouncing off the substrate and returning to the child's hand. If the ball failed to bounce, or if it bounced in such a direction that it failed to return to the child's hand, the event was scored as a failure. By these criteria, bouncing the ball under the leg was the more challenging of the two tests, and was never completed successfully (Fig. 9.7). During the course of a play session, the child switched to the more difficult test on several occasions after repeated success at the less difficult test. Each time, she was unable to complete the more difficult maneuver and switched back to the original test (Fig. 9.8).

In social forms of play, this process of switching among tests of varying difficulty can be viewed in terms of choice of play partners. As long as the win–loss ratio stays within some theoretically definable bounds, play

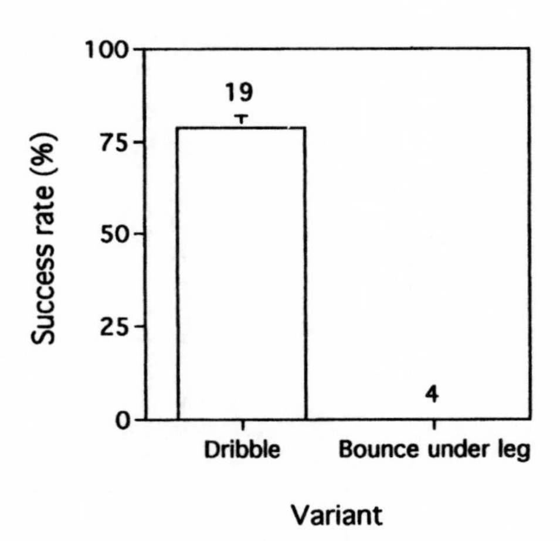

Fig. 9.7. Success rates for two tests performed by a pre-school (four-year old) child with a basketball: dribbling the ball against the ground in the conventional manner and attempting to bounce the ball under a raised leg. Success was defined as an uninterrupted sequence of movement of the ball from the subject's hand to the ground and back to the hand. Numbers above bars indicate the number of times each test was observed.

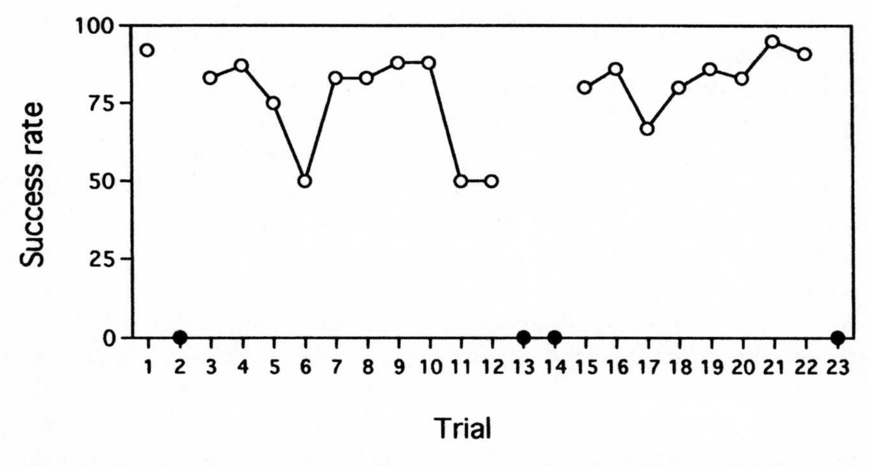

Fig. 9.8. Chronological sequence of events during the play session depicted in Fig. 9.7. Each trial denotes an attempt to bounce the ball in one of the two previously described manners. (o) Dribble, (•) Bounce under leg.

partner preferences should remain intact. Pairs of individuals with more one-sided relationships are less likely to play together. If the play interactions of once well-matched partners become increasingly one-sided, a shift in preferred partners is predicted.

Social play is progressive in that play partners are typically also developing individuals. Play partners that are well-matched (i.e., of similar developmental state) at any given point in time are likely to remain well-matched as long as they continuing to develop at approximately the same rate. This suggests that young individuals should exhibit clear play partner preferences and that these preferences should be highly stable over the course of development.

This is indeed the case for the only species examined to date, the sable antelope (Thompson 1996a). Play partner preferences were related to the age difference between calves, with calves closest in age more strongly preferred. Play partner preferences developed within the first few weeks of life and remained stable for at least several months, during the time when social play is most frequent. Patterns of play bout initiations suggested that calves expanded their pool of play partners in early ontogeny, then concentrated play bout initiations on a subset of preferred peers. Calves born late in the season arrived at their choice of preferred play partners quickly, while calves born early in the season continued to expand their pool of play partners by initiating a few play bouts with the calves born later in the season. This pattern might be expected if calves periodically sample other potential play partners in search of the most suitable partner.

Another aspect of the progressive nature of play, as it relates to self assessment, is the tendency for play to become increasingly social as development progresses. In many species manipulative and locomotor play predominate during early life (West 1974, Baldwin 1986, Chalmers 1980; Gomendio, 1988; Thompson, unpublished data). More interactive forms of play (e.g., play fighting) usually arise later, perhaps after the requisite physical skills have been developed through solitary manipulative and locomotor play. A closer examination of manipulative and locomotor play reveals that they too often become increasingly social over the course of development. Infant antelopes, for example, run by themselves when only days old but quickly shift to running in groups (Thompson, unpublished data). Where solitary runs consisted of chains of successive locomotor movements (running, jumping, kicking), the social runs consist of racing, chasing and dodging. When running in groups calves rarely insert other locomotor behaviors that are commonly observed during solitary runs. This may represent a shift from solitary forms of assessment (success/failure) to social forms (win/loss).

Similar trends are evident in the play of young children. One form of playground play that clearly exemplifies this is the game of 'follow-the-leader.' One child (the leader) assumes a position at the front of a small group and leads the group through a series of obstacles. Each child attempts to traverse the obstacles in turn and generally maintains a fixed position within the group. This is consistent with an interpretation of self assessment of locomotor development through social comparison. Children further engage in social comparison through copying the locomotor activities of others and verbally encouraging others to copy them (e.g., boasting or daring). Self assessment theory leads to the prediction that children who 'lose' during follow-the-leader games (by failing to traverse an obstacle or by falling well behind the group) are likely to return later to the obstacles that they had difficulty with and practice these tests by themselves before engaging in another follow-the-leader episode.

Accuracy of feedback derived from play

Two characteristics of play are crucial to the accuracy of feedback derived from play. First, play fighting should be dyadic. Play bouts involving more than two individuals make determination of winners and losers problematic. If multiple individuals are involved, the bout should take the form of sequential interactions between different pairs of individuals. Second, play should occur at or near maximal intensity (i.e., individuals should 'play to win'). If the effort exerted in play is submaximal, the ability to gauge competence accurately is lost. The logical extension of this idea is that self assessment is incompatible with self handicapping. A 'win' against an opponent that is not actively participating in the interaction provides no useful feedback regarding the 'winner's' competitive ability. Similarly, if partners are not well-matched, it is impossible to discriminate whether a shift in the proportion of wins against a given partner is due to a change in their relative competitive abilities or in the degree of self handicapping by the more skilled of the pair.

Self assessment and creativity

Self assessment theory may provide a framework for the speculative link between play and creativity (Fagen 1981). Self assessment depends upon testing the limits of physical competence. As an individual becomes increasingly competent at a particular test, novel tests of greater difficulty must be generated. Since switching to more challenging tests involves escalation of risk, it is critical that all relevant physical parameters be tested thoroughly prior to attempting a test that would expose the individual to greatly increased risk. In other words, the more well-rounded the feedback, the more informed the decision. Therefore, tests should encompass a wide range of situations that are unfamiliar and unanticipated. Novelty in play would further be favored by natural selection if an individual benefits from being able to assess the full range of competencies prior to their actual use in a fitness-related context.

Avenues for future research

Consideration of self assessment as a mechanism for managing development suggests some particularly fruitful avenues for future research. These include:

1. Testing the relationship between the probability of success and the choice of subsequent tests in solitary object and locomotor play

The sequence of tests should be determined by the pattern of successes and failures (Fig. 9.6). As the probability of success for a given test improves to a level above some threshold, the individual should switch to a more challenging test (i.e., one with a lower probability of success). Conversely, if the probability of success falls below a second threshold, the individual should switch to a test whose probability of success falls above the threshold. Probabilities of success for specific tests are readily quantifiable and should allow determination of the thresholds necessary for switching to more or less challenging tests.

2. Testing the relationship between probability of winning in social play and choice of play partners

Patterns of wins and losses should affect play frequency and partner preference in a manner analogous to the decision rules for solitary play. Preferred play partners should be those that are evenly matched (i.e., those with roughly comparable probabilities of winning). Shifts in the probability of winning against a particular partner should be accompanied by changes in play partner preferences. Individuals that suddenly find themselves losing much more frequently against a formerly well-matched partner should switch to playing with one with whom they have a higher probability of winning and vice versa.

3. Investigating what happens to individuals that can not maintain a win/loss ratio within the apparent bounds necessary to sustain play

The above scenario assumes that individuals that rarely win (or always win) with one particular partner have access to more well-matched partners. In many situations this will not be the case. If an ideal partner is not available, social play may be maintained through self handicapping on the part of the likely winner or by changing the nature of play such that the probability of winning becomes more equal. In the absence of self handicapping by the likely winner, chronic losers may refuse to play.

4. Investigating the precise relationship between risk and play

Because feedback in play comes from monitoring the success of various tests, each test necessarily involves risk of failure. Frequently in locomotor play and potentially also in play with objects, failure may entail substantial risk of injury. Risk of injury may also be a factor in social play bouts, if an individual is greatly over-matched. This aspect of self assessment may lend itself to manipulative experiments in which the degree of risk is experimentally modified.

5. Investigating the role of social comparison in forms of play traditionally considered non-social (object play, group locomotor-rotational play)

The potential for winning and losing in locomotor and object play has not previously been examined. There is much to be learned about the dynamics of group locomotor play, for example. Are start and stop times of individuals within a group locomotor play independent? Are play partner preferences exhibited? What factors determine the onset and termination of group play?

Conclusions

Whatever the ultimate function(s) of play (and there may be many), repeated performance of play behaviors *ad infinitum* would be arguably disadvantageous for several reasons. First, energy allocated to play could be used to increase growth or stored for future needs. These energy costs are typically small (e.g., 2–15% of daily energy expenditure; Martin 1982, 1984; Miller & Byers 1991; Siviy & Atrens 1992) but are possibly non-trivial to a rapidly growing infant. Miller & Byers (1991) estimated that running play represented about 20% of total daily energy expenditure excluding growth for pronghorn fawns, *Antilocapra americana*. Second, play increases exposure to predation, accidents and natural calamities (Fagen 1981; Trillmich 1981; Byers 1987; Harcourt 1991). Additionally, there is the possibility that a threshold level of play is necessary before any benefits accrue, or that any additional play above a certain amount does not result in increased benefits (Martin & Caro 1985). Finally, different types of play and play with different partners probably vary in their effectiveness (Fagen 1981). All these factors may vary with prevailing environmental and social conditions. Thus, the amount of play and the extent of benefits are probably not related in any simple way.

Self assessment provides a means for a developing individual to evaluate the effectiveness of its play experiences and modify its play for optimal benefit while minimizing energy expenditure and exposure to predators. Therefore, self assessment in play could have evolved concurrently with other hypothesized functions, such as the development of predatory and competitive social skills. It is also possible that self assessment has become decoupled from other play functions in certain situations. For example, play fighting may serve as self assessment of competitive capabilities that may influence decisions as varied as whether to try to steal milk from the mother of a play partner, when to disperse from the natal group and whether to challenge another individual for access to mates. Such self assessment of fighting abilities has been proposed for adult red-necked wallabies (Watson & Croft 1993) and may explain the sparring commonly seen in bachelor herds of ungulates.

If animals are using self assessment to manage their play, this might help explain some of the difficulties in studying play. Given self assessment during play, there need be no straightforward relationship between play content and frequency under self-selected conditions and predicted outcome variables such as predatory and fighting skill. Self assessment provides a mechanism of self correction and a means of buffering development against perturbations in the physical and social environment. This capability for developmental flexibility might be extremely important for species in which the social or physical environment is unpredictable or unstable. Likewise it would be beneficial for species in which developmental perturbations and/or experiences have long-lasting effects. Rather than increasing physical capabilities or speeding up the acquisition of skills, play may simply ensure that development proceeds normally in differing (and potentially unpredictable) environments. In this context, it is interesting that play seems extremely sensitive to subtle perturbations in the physical and social environment (Thompson 1996b).

Acknowledgments

I am indebted to all who helped and inspired me during my graduate and post-graduate research on play behavior. First, I am grateful to Jack Cranford for permission to study the punarés at Virginia Polytechnic Institute and State University and for his mentorship during my graduate work there. I also thank my doctoral dissertation advisor, Devra Kleiman for her feedback and encouragement. I thank Chris Wemmer and Larry Collins for permission to study the ungulates at the National Zoological Park's Conservation and Research Center. Support was provided by the Department of Biology, V.P.I. & S.U., the Department of Zoology and College of Life Sciences, University of Maryland, a Smithsonian Institution Graduate Fellowship and a National Science Foundation Graduate Fellowship. Lastly, I would like to acknowledge the cooperation of the Center for Young Children at the University of Maryland, College Park and the assistance of Eduardo Ruiz, Michael

McIlwain, and Gjange Smith with my study of play behavior in pre-school children.

References

Aldis, O. 1975. *Play Fighting*. New York: Academic Press.

Baldwin, J. D. 1986. Behavior in infancy: exploration and play. *Primate Biology, Vol. 2A: Behavior, Conservation, and Ecology*, pp. 295–326. New York: Alan R. Liss.

Bateson, P. P. G., Mendl, M., & Feaver, J. 1990. Play in the domestic cat is enhanced by rationing of the mother during lactation. *Anim. Behav.*, **40**, 514–25.

Bekoff, M. 1974. Social play and play-soliciting by infant canids. *Am. Zool.*, **14**, 323–40.

Berger, J. 1980. The ecology, structure and function of social play in Bighorn sheep (*Ovis canadensis*). *J. Zool. Lond.*, **192**, 531–42.

Bernstein, I. 1981. Dominance: the baby and the bathwater. *Behav. Br. Sci.*, **4**, 419–57.

Biben, M. 1982. Object play and social treatment of prey in bush dog and crab-eating foxes. *Behaviour*, **79**, 201–11.

Biben, M. 1983. Comparative ontogeny of social behavior in three South American canids, the maned wolf, crab-eating fox and bush dog: implications for sociality. *Anim. Behav.*, **31**, 814–26.

Biben, M. 1986. Individual- and sex-related strategies of wrestling play in captive squirrel monkeys. *Ethol.*, **71**, 229–41.

Biben, M. 1989. Effects of social environment on play in squirrel monkeys: resolving Harlequin's dilemma. *Ethol.*, **81**, 72–82.

Boulton, M. J. 1991. Partner preferences in middle school children's playful fighting and chasing: a test of some competing functional hypotheses. *Ethol. Sociobiol.*, **12**, 177–93.

Brueggeman, J.A. 1978. The function of adult play in free-ranging *Macaca mulatta*. In: *Social Play in Primates* (ed. E. O. Smith), pp. 169–92. New York: Academic Press.

Byers, J. A. 1980. Play partner preferences in Siberian ibex, *Capra ibex siberica. Z. Tierpsychol.*, **53**, 23–40.

Byers, J. A. 1984. Ungulate play behavior. In: *Play in Animal and Humans* (ed. P. K. Smith), pp. 43–65. New York: Blackwell.

Byers, J. A. 1987. Why the deer and the antelope play. *Natural History*, May, 54–61.

Byers, J. A. & Walker, C. 1995. Refining the motor training hypothesis for the evolution of play. *Am. Nat.*, **146**, 25–40.

Caro, T. M. 1988. Adaptive significance of play: are we getting closer? *Trends Ecol. Evol.*, **3**, 50–4.

Caro, T. M. 1995. Short-term costs and correlates of play in cheetahs. *Anim. Behav.*, **49**, 333–45.

Carpenter, C. R. 1934. A field study of the behavior and social relations of howling monkeys. *Comp. Psych. Mono.*, **10**, 1–168.

Chalmers, N. R. 1980. The ontogeny of play in feral olive baboons (*Papio anubis*). *Anim. Behav.*, **29**, 570–95.

Dane, B. 1977. Mountain goat social behavior: social structure and 'play' behavior as affected by dominance. In: *Proceedings of the First International Mountain Goat Symposium* (eds. W. M. Samuel & W. G. Macgregor), pp. 92–106. British Columbia: Ministry of Recreation and Conservation.

Drea, C. M., Hawk, J. E., & Glickman, S. E. 1996. Aggression decreases as play emerges in infant spotted hyenas: preparation for joining the clan. *Anim. Behav.*, **51**, 1323–36.

Fagen, R. M. 1976. Exercise, play, and physical training in animals. In: *Perspectives in Ethology, Vol. 2* (eds. P. P. G. Bateson & P. H. Klopfer), pp. 189–219. New York: Plenum Press.

Fagen, R. 1981. *Animal Play Behavior*. New York: Oxford University Press.

Fagen, R. M. & George, T. K. 1977. Play behavior and exercise in young ponies (*Equus caballus* L.). *Behav. Ecol. Sociobiol.*, **2**, 267–9.

Fry, D. P. 1987. Differences between playfighting and serious fighting among Zapotec children. *Ethol. Sociobiol.*, **8**, 285–306.

Gomendio, M. 1988. The development of different types of play in gazelles: implications for the nature and functions of play. *Anim. Behav.*, **36**, 825–36.

Groos, K. 1898. *The Play of Animals* (translated by E. L. Baldwin). London: D. Appleton & Co.

Harcourt, R. 1991. Survivorship costs of play in the South American fur seal. *Anim. Behav.*, **42**, 509–11.

Havkin, Z. & Fentress, J. C. 1985. Forms of combative strategy in interactions among wolf pups (*Canis lupus*). *Z. Tierpsychol.*, **68**, 177–200.

Hole, G. 1988. Temporal features of social play in the laboratory rat. *Ethol.*, **78**, 1–20.

Jamieson, S. H. & Armitage, K. B. 1987. Sex differences in the play behavior of yearling yellow-bellied marmots. *Ethol.*, **74**, 237–53.

Loizos, C. 1966. Play in mammals. *Symp. Zool. Soc. Lond.*, **18**, 1–9.

Martin, P. 1982. The energy cost of play: definition and estimation. *Anim. Behav.*, **30**, 294–5.

Martin, P. 1984. The (four) whys and wherefores of play in cats: a review of functional, evolutionary, developmental and causal issues. In: *Play in Animals and Humans* (ed. P.K. Smith), pp. 159–73. New York: Blackwell.

Martin, P. & Caro, T. M. 1985. On the functions of play and its role in behavioral development. *Adv. Study Behav.*, **15**, 59–103.

McDonald, D. L. 1977. Play and exercise in the California ground squirrel, (*Spermophilus beecheyi*). *Anim. Behav.*, **25**, 782–6.

Miller, M. N. & Byers, J. A. 1991. Energetic costs of locomotor play in pronghorn fawns. *Anim. Behav.*, **41**, 1007–13.

Moore, C. L. & Power, K. L. 1992. Variation in maternal care and individual differences in play, exploration and grooming of juvenile Norway rat offspring. *Dev. Psychobiol.*, **25**, 165–82.

Owens, N. W. 1975a. Social play behaviour in free-living baboons, *Papio anubis*. *Anim. Behav.*, **23**, 387–408.

Owens, N. W. 1975b. A comparison of aggressive play and aggression in free-living baboons, *Papio anubis*. *Anim. Behav.*, **23**, 757–65.

Panksepp, J. 1981. The ontogeny of play in rats. *Dev. Psychobiol.*, **14**, 327–32.

Pellis, S. M. 1988. Agonistic versus amicable targets of attack and defense: consequences for the origin, function and descriptive classification of play-fighting. *Aggress. Behav.*, **14**, 85–104.

Pellis, S. M. 1993. Sex and the evolution of play fighting: a review and model based on the behavior of muroid rodents. *Play Theory Res.*, **1**, 55–75.

Pellis, S. M. 1991. How motivationally distinct is play? A preliminary case study. *Anim. Behav.*, **42**, 851–3.

Pellis, S. M. & Pellis, V. C. 1983. Locomotor-rotational movements in the ontogeny and play of the laboratory rat *Rattus norvegicus. Dev. Psychobiol.*, **16**, 269–86.

Pellis, S. M. & Pellis, V. C. 1988. Play-fighting in the Syrian golden hamster *Mesocricetus auratus* Waterhouse, and its relationship to serious fighting during postweaning development. *Dev Psychobiol.*, **21**, 333–7.

Poole, T. B. 1966. Aggressive play in polecats. *Symp. Zool. Soc. Lond.*, **18**, 23–44.

Renouf, D. & Lawson, J. W. 1986. Play in fur seals (*Phoca vitulina*). *J. Zool. Lond.*, **208**, 73–82.

Rowell, T. E. 1973. *The Social Behavior of Monkeys*. Baltimore, Maryland: Penguin Books.

Siviy S. M. & Atrens, D. M. 1992. The energetic costs of rough-and-tumble play in the juvenile rat. *Dev. Psychobiol.*, **25**, 137–48.

Smith, E. F. S. 1991. The influence of nutrition and postpartum mating on weaning and subsequent play behaviour of hooded rats. *Anim. Behav.*, **41**, 513–24.

Smith, P. K. 1982. Does play matter? Functional and evolutionary aspects of animal and human play. *Behav. Brain Sci.*, 5, 139–84.

Steiner, A. L. 1971. Play activity of Columbian ground squirrels. *Z. Tierpsychol.*, **28**, 247–61.

Stevenson, M. F. & Poole, T. B. 1982. Playful interactions in family groups of the common marmoset (*Callithrix jacchus jacchus*). *Anim. Behav.*, **30**, 886–900.

Symons, D. 1978. *Play and Aggression: A Study of Rhesus Monkeys*. New York: Columbia University Press.

Terranova, M. L. & Laviola, G. 1995. Individual differences in mouse behavioural development: effects of precious weaning and ongoing sexual segregation. *Anim. Behav.*, **50**, 1261–71.

Thompson, K. V. 1985. Social play in the South American punaré, (*Thrichomys apereoides*): a test of play function hypotheses. M.Sc. thesis, Virginia Polytechnic Institute and State University.

Thompson, K. V. 1996a. Play-partner preferences and the function of social play in infant sable antelope, *Hippotragus niger*. *Anim. Behav.*, **52**, 1143–55.

Thompson, K. V. 1996b. Behavioral development and play. In: *Wild Mammals in Captivity: Principles and Techniques* (eds. D. G. Kleiman, M. E. Allen, K. V. Thompson, S. Lumpkin & H. Harris), pp. 352–71. Chicago: University of Chicago Press.

Trillmich, F. 1981. Mutual mother-pup recognition in Galapagos fur seals and sea lions: cues and functional significance. *Behaviour*, **78**, 21–42.

Watson, D. M. 1993. The play associations of red-necked wallabies. *Ethol.*, **94**, 1–20.

Watson, D. M. & Croft. D. B. 1993. Playfighting in captive red-necked wallabies, *Macropus rufogriseus banksiansus*: *Behaviour.*, **127**, 219–45.

Watson, D. M. & Croft, D. B. 1996. Age-related differences in playfighting strategies of captive male red-necked wallabies (*Macropus rufogriseus banksianus*). *Ethol.*, **102**, 336–46.

West, M. 1974. Social play in the domestic cat. *Am. Zool.*, **14**, 427–36.
Wilson, S. & Kleiman, D. 1974. Eliciting play: a comparative study. *Am. Zool.*, **14**, 331–70.

10

Biological effects of locomotor play: getting into shape, or something more specific?

JOHN A. BYERS
Department of Biological Sciences, University of Idaho, Moscow, ID 83844–3051 USA

Moving the skeleton is an engineer's nightmare.

(Thach 1996, p. 415)

Arnold Schwarzenegger shows us one side of the profound plasticity of the vertebrate muscular–skeletal system. Muscles become larger and bones appropriately remodeled when they experience increased work loads. Parenthetically, although Mr. Schwarzenegger has not been called upon to go topless in his recent films, his continuing apparent bulk under a sports jacket shows that use-specific hypertrophy is not confined to a narrow age range. I shall return to this point later. The anti-Arnold effect, or disuse-specific atrophy, also is well known. It is strikingly illustrated by the rapid loss of muscle and bone mass that occurs in zero gravity. Mammals, including humans, that spend a few days in earth orbit return with substantial reduction in muscle and bone mass (Bodine-Fowler et al. 1992; Cann & Adachi 1983; Morey & Baylink 1978). NASA acknowledges that muscular–skeletal atrophy is the most serious medical problem associated with space flight. Another well-known plastic response in vertebrates is the use-induced gain and disuse-induced loss of aerobic capacity. Here, changes in many organ and enzyme systems are involved (Close 1972; Laughlin et al. 1989; Nieman 1990; Bigard et al. 1991), and the effects are equally as dramatic as Mr. Schwarzenegger's pectoralis muscles. I and most other humans cannot run a single 5 minute mile, but some humans who have trained for marathons can run 26 of these in succession.

Phenotypic plasticity in systems that support strength and endurance is adaptive. The muscular–skeletal system has a way of detecting demands placed upon it, and responds appropriately, building as much, but no more, capability as is called for. In the economy of nature, a capability called for but not met equals death; a capability in excess equals squandered opportunity to reproduce. Therefore, it is not surprising that phenotypic plasticity is widespread in nature. Vertebrate muscular–skeletal plasticity is by no means unique. Phenotypic plasticity is the norm rather than the exception in plants (Sultan 1987; Karban & Myers 1989) and animals (Palumbi 1984; Meyer 1987; Metcalfe et al. 1989; West-Eberhard 1989; Bronmark & Miner 1992).

The joint results of muscular–skeletal remodeling and increase in aerobic capacity that accompany exercise are commonly known as 'being in shape.' The observation that most locomotor and social play of young mammals is exercise has naturally prompted several versions of what I call the 'getting into shape' hypothesis for the function of play which proposes that a main function of play is to provide 'exercise,' (Fagen 1976) or 'physical training' (Fagen 1981; Bekoff 1988). In many published descriptions of play, this getting-into-shape hypothesis is mentioned and casually accepted, as if it is obvious that getting into shape is a biological consequence of play (Fairbanks 1993; Biben, Chapter 8). However, the fact that play provides exercise is not in itself sufficient proof that play gets animals into shape. Not just any amount or kind of exercise will prompt a training response. Daily tooth brushing provides some exercise to the deltoids, but even the most assiduous program of dental hygiene will not create deltoids like those of Mr. Schwarzenegger.

In this chapter, I show that the getting-into-shape hypothesis is unlikely, for three reasons. First, all getting-into-shape physiological responses are very transitory, and disappear shortly after exercise stops. Because play in most species occurs only during a limited period of juvenile life, its getting-into-shape effects, if they existed, would not persist. Thus one would have to find a getting-into-shape advantage for juveniles. At this time, no such advantage has been shown, and, in fact it seems that play has small survival costs, not benefits. Second, the kind and amount of exercise performed in play in most species is probably insufficient to prompt physiological training responses. Third, exercise responses are not age-limited; mammals can exercise and obtain a training response at any age. If play was designed as getting-into-shape exercise, I would expect it to be scheduled at many different ages in different species, depending on specific life histories. Instead, play in most mammals occurs only in a narrow range of ages early in life. Finally in this chapter, I suggest a method by which one can evaluate any other functional hypothesis (e.g., play provides motor learning) about play. The method starts from the premise that the age distribution of play is an essential clue about play's true biological effects. Effects that are permanent, and available only during ages when play typically occurs, are likely true effects.

The earliest of the getting-into-shape hypotheses was that of Brownlee (1954), who proposed that, during infancy, some muscles would not receive use in normal activity sufficient to prevent their atrophy. In other words, Brownlee proposed that there are 'play muscles,' that will be used in adult life, but which are used seldom, if at all, in juvenile life. In Brownlee's view, play exists to work those muscles in juvenile life, to prevent their atrophy. In my opinion, Brownlee deserves a lot of credit for the first clear articulation of a testable hypothesis for the function of play. Brownlee's hypothesis predicts: (1) that some muscles used in adult life receive little or no use in juvenile life, (2) that these muscles, if not worked in juvenile life, will be incapable of any adaptive response to the work loads that they receive in adult life. To my knowledge, no one has tested either of these hypotheses. Although a test of Brownlee's hypothesis might yield some interesting surprises, I think that it, as well as the more general getting-into-shape hypotheses are unlikely. I say this because most play ends before adult life begins. Usually, it ends long before adult life begins.

The age distribution of play is neglected evidence

Consider the age distributions of play shown in Fig. 10.1, which depicts ontogeny of play for an ungulate, a rodent, a primate, and a carnivore. There are three points to notice. First, play does not occur immediately after birth; it appears sometime later. Second, after play appears, its rate of expression rises quickly to a peak. Third, the peak of expression of play is brief, compared to the life span of the species. Rate of play falls quickly to zero or near zero. Play may appear sporadically later in life, but usually unpredictably and at a low rate, compared to peak rates in juvenile life. In summary, data in Fig. 10.1, and from every other study I know of in which age-specific rates of play are given, show that the expression of play is discrete. Play is turned on, then turned off. A switch is the correct metaphor for many species such as house mice or fat-tailed dunnarts, but for other species, a rheostat is a better metaphor: play is turned on, then rapidly dimmed. To me, such a tight age distribution has but one plausible explanation: play must represent a sensitive period (Immelman & Suomi 1981) during which the performance of certain

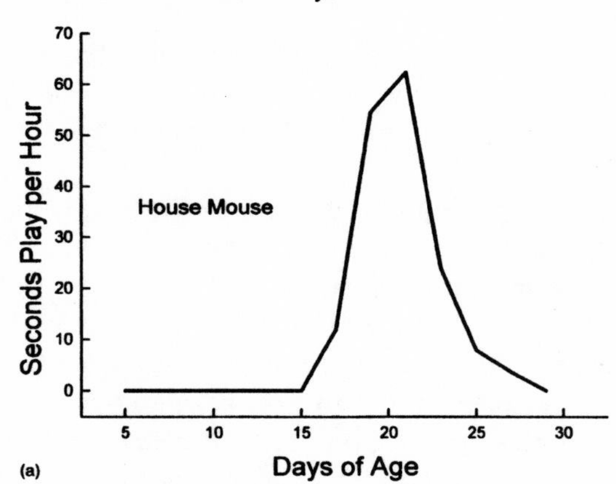

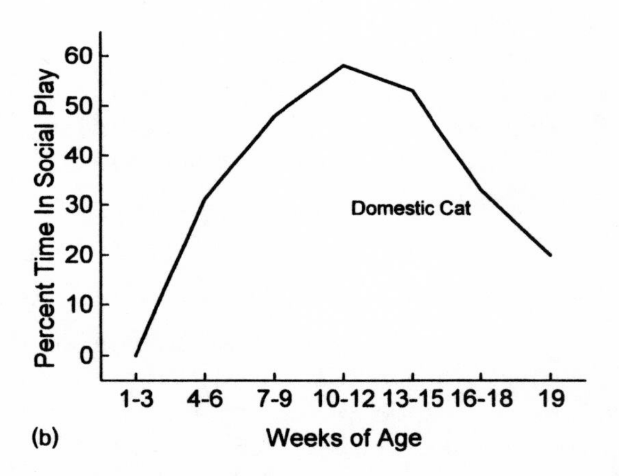

Fig. 10.1. The age distribution of play rate in four species of mammal. Data redrawn from the following sources: (*a*) House mouse, Byers and Walker (1995); (*b*) domestic cat, West (1974); (*c*) pronghorn, Byers (1997); (*d*) Olive baboon, Owens (1975).

motor patterns can alter development. A well known example of such development occurs in many songbirds, in which a young bird must perform the motor act of singing (subsong) to modify structure of the neurons that issue motor commands for singing (Marler 1970; Kroodsma 1981).

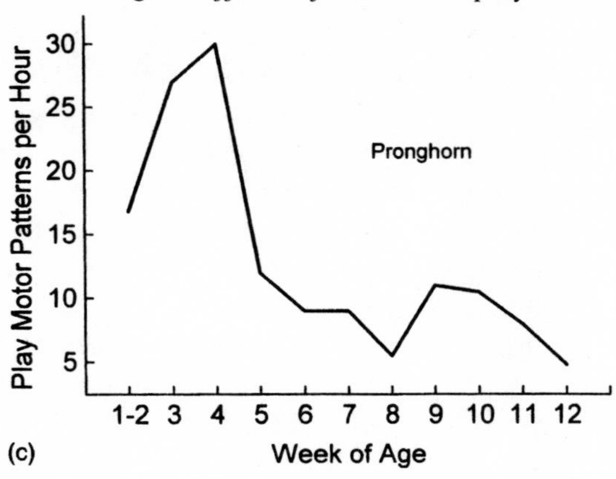

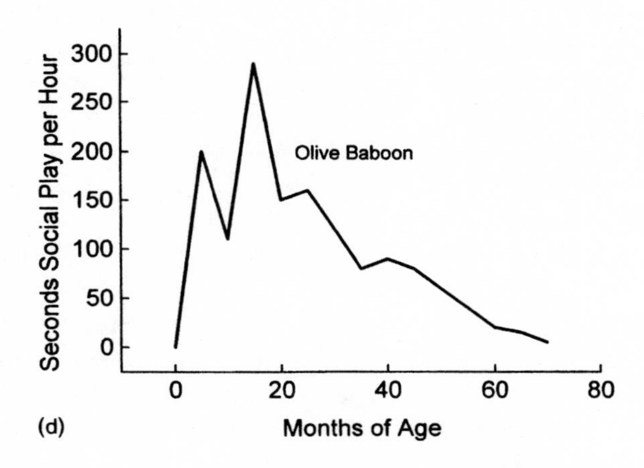

Fig. 10.1 (*cont.*)

Alternative hypotheses to explain the age distribution of play

One alternative to my interpretation is that the age distribution has been created by selection for life history optimality. This approach (Fagen 1977, 1981) uses the following premise and question. If the beneficial effects of play are permanent, what would be the optimal age expression of play? The models predict that play should occur early in life, but they do not accurately predict the ages of onset and of termination. Also, the models require the benefits of play to be permanent: 'The model just discussed demonstrates that the hypothesized delayed benefits of play are probably feasible in an evolutionary sense. Physical training and skill development effects provide the necessary mechanisms for such delayed benefits.' (Fagen 1981, p. 365). However, as I shall show, most physical training effects are *not* permanent – they disappear quickly after exercise stops. Therefore, I conclude that Fagen's life history models are not a good explanation for the typical age distribution of play.

A third explanation of play's age distribution is that its effects are immediate and transitory (Martin & Caro 1985). Perhaps play exists solely because it increases survival of juveniles. This view is difficult to reconcile with findings that play has substantial energy costs (Martin 1984, Miller & Byers 1991, Siviy & Atrens 1992) as well as survival costs (Welles & Welles 1961; Byers 1977; Berger 1979; 1980; Caro 1987; 1994; Harcourt 1991). I saw several Siberian ibex kids injured in the course of their acrobatic play, which included running and leaping on steep slopes (Byers 1977). Berger (1979) saw bighorn sheep lambs injured on cactus and Welles & Welles (1961) saw lambs fall to their death in play. Caro (1987) found that cheetah cub play disrupted the hunting success of mothers. Eighty-five percent of South American fur seal pups that Harcourt (1991) saw killed by Southern sea lions were taken during play in shallow water. I am aware of no study that has shown that play increases juvenile survival. Rather, the data (admittedly from a few studies) suggest that survivorship costs of play are neutral (Byers 1997) or slightly positive.

In summary, there are three possible explanations for the typical age distribution of play: (1) sensitive period; (2) delayed permanent effects; (3) immediate increase in juvenile survival. Hypothesis 2 is suspect because most physical training effects are transitory. Hypothesis 3 is implausible because play seems to have mild survival costs, not benefits. Hypothesis 1 thus becomes attractive by process of elimination. Also, as I shall show, there is other compelling evidence for the sensitive period hypothesis.

The sensitive period hypothesis

A sensitive period in behavioral development refers to a window in development during which specific types of experience permanently alter the course of development of the brain or of other systems that support behavior. The experience-dependent development is not possible at ages before or after the window. Sensitive periods are common. Many songbirds must hear species-specific song early in life to properly develop singing (Kroodsma 1981). A human girl deprived of language experience in childhood developed only rudimentary speech (Curtiss 1977), human phonetic perception of speech sounds is modified by exposure to speech before six months of age (Kuhl et al. 1992), and human ability to master a second language is strongly dependent on exposure to that language before about seven years of age (Locke 1993). In many species of precocial birds, young show a short period of irreversible learning in which they develop a following preference for the parent (Lorenz 1935; Hess 1973). Many birds show a longer and later period during which they develop identification mechanisms that promote reproductive isolation (Immelmann 1972; Cooke & McNally 1975). Water stick insects (*Ranatra linearis*) after each molt perform a series of slow motion prey capture movements in which they calibrate the strike to the new size of the forelegs (Cloarec 1982). Lobsters (*Homarus americanus*) normally develop one cutter and one crusher claw. If individuals are deprived of small objects to manipulate in the juvenile fourth and fifth stages, they develop symmetrical cutter claws (Govind 1989). Young pigeons and probably young of other migratory birds develop a sun compass only during a discrete period of juvenile life, and must see the sun while they are flying to do so (Keeton 1981). Among mammals, there is a kind of global sensitivity to early experience. Experimental or naturally produced differences in the early experience of mammals have diverse effects on many dimensions of their behavior (Bekoff 1977; Einon at al. 1977; Greenough & Juraska 1979; Novakova & Babicky 1979; Adamec et al. 1980; Huck & Banks 1980; Chivers & Einon 1982; Fillion & Blass 1986; Bekoff 1989a; Byers 1997). Because sensitive periods are common, it is reasonable to postulate that play, with its discrete age range of expression, may be another example of performance-dependent development. But what corroborative evidence is there to support such a claim?

Evidence for the sensitive period hypothesis

The sensitive period hypothesis came to me in an instant one afternoon several years ago in the University of Idaho library. I was idly looking through some books on the cerebellum when I came across Larramendi's (1969) Figure 9 that showed the postnatal development of parallel fiber synapses on Purkinje dendrites in the mouse cerebellum. I realized, looking at Larramendi's figure, that I might as well be looking at a depiction of the age distribution of play in mice. I then realized that I had an explanation for the age distribution of play. Play occurred during the ages when it was possible for motor activity to alter the terminal phase of synapse formation and elimination in the area of the brain that controls coordinated motor output. That cerebellar synaptogenesis can be influenced by early rearing environment was already well known (Greenough & Juraska 1979; Brown et al. 1991). This preliminary evidence for a specific biological effect of play led me to question whether the broad 'getting-into-shape' hypothesis could be correct and led me and Curt Walker (Byers & Walker 1995) to perform the following analysis.

We started with the premise that the age distribution of play is evidence about its biological effects. We surveyed the exercise physiology literature, examining a number of specific exercise effects, and asked two questions about each. First, is the effect permanent? Second, is the effect age-dependent? We reasoned that the 'sensitive period' hypothesis would become more plausible if we could find exercise effects that were permanent, and were available only during post natal ages when play typically occurs. We found that none of the getting-into-shape effects met the criteria of permanence and age-dependence, but that two effects associated with skill or economy of movement did.

Why play is not getting into shape

We (Byers & Walker 1995) surveyed 16 specific exercise effects that are major contributors to the physiological training response, which is commonly known as 'being in shape'. The effects, such as VO_2 Max., increase in blood volume, bone remodeling and increased muscle fiber diameter, all support the development of increased strength and endurance in response to exercise. None of these effects is permanent. In fact, all are quite fleeting. The increased capacity for activity created by exercise persists only while the exercise persists. Any athlete or exercise enthusiast will verify these findings. This sort of finely-tuned, reversible plasticity reflects the economy of nature that I mentioned earlier. Organisms rarely retain excess capacity when the materials and energy involved can be shunted to reproduction. The rapid reversibility of the effects of getting-into-shape means that these can be considered at best as immediate, transitory benefits of play. Anyone who wants to hold to the getting-into-shape hypothesis needs to show that such fleeting increases in physical conditioning create a survival advantage. As I noted earlier, evidence to date suggests that play has slight survival costs, not benefits.

Another problem for the getting-into-shape hypothesis is that the level of exercise performed in the play of many species may not be sufficient to evoke the physiological training responses that constitute getting-into-shape. In other words, although locomotor rotational and much social play is exercise, it may not be a good workout. Locomotor play bouts of mice (Walker & Byers 1991) and social play bouts of Norway rats (Birke & Sadler 1984) usually last less than 20 seconds. In contrast, laboratory studies of the exercise training response in these species usually exercise animals for one hour or to exhaustion (Xia 1990; Picarro et al. 1991). Anyone who wants to claim that play gets animals into shape should first show that the level of exercise performed in play is adequate to promote physiological training responses either aerobically or anaerobically (Bekoff 1989b). I predict that in most, if not all species, the true work load of play will not meet this criterion.

A final problem with the getting-into-shape hypothesis is that animals can get into shape by exercising at any age. All of the specific training responses that Byers and Walker (1995) surveyed showed no age limitation. These findings are shown in effective summary form by Arnold Schwarzenegger's continuing bulk beneath a sports jacket, the graying Carl Lewis's 1996 Olympic long jump gold medal, and by Fig. 10.2, which depicts levels of athletic performance achieved by men of different ages. Note that 40 to 50 year-old men, at the historical age of death in humans, can train and perform at better than 90% of Olympic levels. If

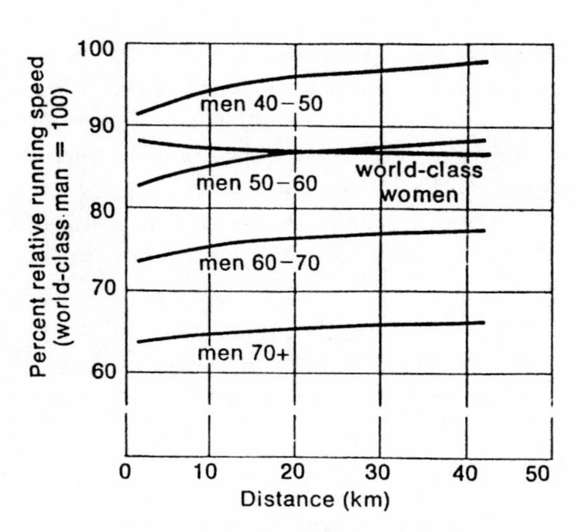

Fig. 10.2. Running speed as a function of race distance in age classes of athletically trained men in compared to Olympic standards in these races. Adapted from Riegel (1981) with permission of the publisher.

getting into shape is transient and also available at any age, why does play have such a tight and narrow age range in most species?

Motor skill and economy effects that may fit the sensitive period hypothesis

In contrast to all strength and economy effects, we (Byers & Walker 1995) found two effects linked to the skill or economy of movement that are permanent and age-dependent. These are: (1) modification of the terminal phase of Purkinje cell synapse formation and elimination and (2) modification of muscle fiber type differentiation.

Cerebellum development

The cerebellum controls the accuracy and form details of movement (Thach et al. 1992; Thach 1996). All output of the cerebellum is inhibitory, by Purkinje cell axons. In young mammals, Purkinje dendrites form many synapses with climbing fibers and parallel fibers after birth. Initially, more synapses are formed than will be retained. Experiments have shown that which synapses are retained, as well as the number of synapses per Purkinje cell, is in part experience-dependent (Greenough & Juraska 1979; Brown et al. 1991). These effects appear to be permanent and they are confined to a narrow age range when final synapse formation and elimination in the cerebellum occurs. Byers & Walker (1995) showed that, in three species (house mice, Norway rats, domestic cats) for which both kinds of data were available, this period of experience-dependent synapse development coincided closely with the age distribution of play. In other words, in these three species, play is turned on when there is the opportunity for experience-dependent modification of the cerebellum, and it is turned off shortly after the architecture of the cerebellum is complete. I think that this is compelling evidence for a specific, sensitive period effect of play. Also, note that the effect, if it exists, is different from motor learning. Motor learning, the acquisition of the ability to perform new skilled motor tasks, can also occur in adult life (Black et al. 1990; Thach 1996)

Muscle fiber type differentiation

Mammalian muscles go through a post-natal phase of development during which the pattern of use of each muscle sets its fiber type composition. This role of motor neuron innervation in setting fiber type composition was discovered serendipitously by the now famous cross innervation experiments of Sir John Eccles and collaborators in 1958 (Buller et al. 1960; Buller & Pope 1977). In two to three week-old cats, Buller et al. surgically attached the motor nerve of the soleus (in the adult a predominately slow twitch muscle) to the flexor digitorum longus (in the adult a predominately fast twitch muscle), and vice-versa. Several months later, the surprise was that the soleus, receiving innervation from a motor nerve that innervates a fast-twitch muscle, had become a fast twitch muscle; the flexor digitorum longus, receiving innervation from a motor nerve that innervates a slow twitch muscle, had become a slow twitch muscle.

Fiber type composition determines the muscle's speed of contraction, maximum force, and the energetic cost of contraction. A muscle's speed of contraction and maximum force are inversely related and are a consequence of its fiber type composition. Slow fibers contract more slowly than fast fibers, but develop greater force. Fast fibers are specialized to contract quickly; they develop less force than slow fibers, and their energetic cost of contraction is greater (Rall 1985). Each muscle in adult life has a characteristic fiber type composition (ratio of and spatial array of fast and slow fibers) that creates the muscle's contraction characteristics (Gans & Gaunt 1991, Gordon & Pattullo 1993). In mammalian muscle, fibers begin development as the slow type; some then differentiate into fast fibers as a consequence of the patterns of innervation that they receive (Armstrong 1980; Baldwin 1984). Thus, the postnatal pattern of use of a muscle can influence its eventual fiber type composition. Eventual fiber type composition also seems to be permanent in natural life, even though the fiber type composition of adult muscle can be experimentally changed by cross innervation or limb suspension (Baldwin 1984; Roy et al. 1991; Gordon & Pattullo 1993).

When we (Byers & Walker 1995) compared the time course of muscle fiber type differentiation to the time course of play in house mice, Norway rats, and domestic cats, we found a result similar to that for cerebellar synaptogenesis. In these three species, the developmental window between the age at which polyneuronal innervation of muscle fibers ends and the age at which muscle fiber type differentiation is complete occurs during the ages when play reaches its peak rates. Another specific biological effect of play may be to modify postnatal muscle fiber type differentiation.

Conclusions

When I looked for data such as those in Fig. 10.1, I was surprised by the scarcity of good descriptive data on play. One might guess that it would be easy to find data on age-specific rates of play in many species. This is not so. Published descriptions of the motor patterns performed in play are abundant, but data on age-specific rates of play are very rare. Thus, while I believe that more studies will show that the time course of play of most mammals is like those shown in Fig. 10.1, I will be happier when the data are in on representatives of many mammal Families. It may be unfashionable to collect descriptive data (at least in behavior: it is still glamorous to do so in molecular biology), but there is still a need for good accounts of the rates of performance of specific motor patterns throughout behavioral development. Perhaps we could borrow a term from the molecular biologists, and say that we are *characterizing* behavioral development. I was astounded to find that there appears to be no comprehensive data on rates of human locomotor play from birth to sexual maturity.

I am harping on the need for good descriptions of development because, as I mentioned earlier, I think that the age distribution of play provides essential clues about the true biological effects of this enigmatic juvenile behavior. The method that I have illustrated here is to look for permanent age-limited effects of repetitive motor performance, and to compare the ages of availability of the effects to the age distribution of play. The suggestive correlations that have emerged from such an analysis do not *prove* biological effects, but they do make specific, testable predictions. In my opinion, the age distribution of play is the key that opens the right doors of investigation.

Acknowledgments

My work in developing these ideas was supported by Public Health Service grant NICHD 22606. I thank Curt Walker for his diligent slogging through a diffuse, frustrating literature.

References

Adamec, R. E., Stark-Adamec, C., & Livingston, K. E. 1980. The development of predatory aggression and defense in the domestic cat (*Felis catus*) I. Effects of early experience on adult patterns of aggression and defense. *Behav. Neural Biol.*, **30**, 389–409.

Armstrong, R. B. 1980. Properties and distributions of the fiber types in the locomotory muscles of mammals. In *Comparative physiology: Primitive mammals.* (eds. K. Schmidt-Nielsen, L. Bolis, C. R. Taylor, P. J. Bentley, & C. E. Stevens) pp. 243–54. Cambridge: Cambridge University Press.

Baldwin, K. M. 1984. Muscle development: neonatal to adult. *Exercise Sports Sci. Rev.*, **12**, 1–20.

Bekoff, M. 1977. Mammalian dispersal and the ontogeny of individual behavioral phenotypes. *Am. Nat.*, **111**, 715–32.

Bekoff, M. 1988. Motor training and physical fitness: possible short- and long-term influences on the development of individual differences in behavior. *Dev. Psychobiol.*, **21**, 601–12.

Bekoff, M. 1989a. Social development of terrestrial carnivores. In *Carnivore Behavior, Ecology, and Evolution*, (ed. J. L. Gittleman), pp. 89–124. Ithaca, NY: Cornell University Press.

Bekoff, M. 1989b. Social play and physical training: when 'not enough' may be plenty. *Ethology*, **80**, 330–3.

Berger, J. 1979. Social ontogeny and behavioural diversity: consequences for bighorn sheep *Ovis canadensis* inhabiting desert and mountain environments. *J. Zool., Lond.* **188**, 251–66.

Berger, J. 1980. The ecology, structure and functions of social play in bighorn sheep (*Ovis canadensis*). *J. Zool., Lond.*, **192**, 531–42.

Bigard, A. X., Brunet, A., Guezennec, C. Y., & Monod, H. 1991. Skeletal muscle changes after endurance training at high altitude. *J. Appl. Physiol.*, **71**, 2114–21.

Birke, L. I. A., & Sadler, D. 1984. Modification of juvenile play and other social behaviour in the rat by neonatal progestins: further studies. *Physiol. Behav.* **33**, 217–19.

Black, J. E., Isaacs, K. R., Anderson, B. J., Alcantara, A. A., & Greenough, W. T. 1990. Learning causes synaptogenesis, whereas motor activity causes angiogenesis, in cerebellar cortex of adult rats. *Proc. Natl. Acad. Sci., USA* **87**, 5568–72.

Bodine-Fowler, S., Roy, R. R., Rudolph, W., Hague, N., Kozlovskaya, I. B., & Edgerton, V. R. 1992. Spaceflight and growth effects on muscle fibers in the rhesus monkey. *J. Appl. Physiol. Supl.*, **73**, S82–S89.

Bronmark, C., & Miner, J. G. 1992. Predator-induced phenotypical change in body morphology in crucian carp. *Science*, **258**, 1348–50.

Brown, M. C., Hopkins, W. G., & Keynes, R. J. 1991. *Essentials of Neural Development*. Cambridge: Cambridge University Press.

Brownlee, A. 1954. Play in domestic cattle in Britain: an analysis of its nature. *Brit. Vet. J.* **110**, 48–68.

Buller, A. J. Eccles, J. C., & Eccles, R. M. 1960. Differentiation of fast and slow muscles in the cat hind limb. *J. Physiol.*, **150**, 399–416.

Buller, A. J., & Pope, R. 1977. Plasticity in mammalian skeletal muscle. *Phil. Trans. Roy. Soc. Lond., B*, **278**, 295–305.

Byers, J. A. 1977. Terrain preferences in the play of Siberian ibex kids (*Capra ibex sibirica*). *Z. Tierpsychol*, **45**, 199–209.

Byers, J. A. 1997. *American Pronghorn. Social Adaptations and the Ghosts of Predators Past*. Chicago: University of Chicago Press.

Byers, J. A., & Walker, C. B. 1995. Refining the motor training hypothesis for the evolution of play. *Am. Nat.*, **146**, 25–40.

Cann, C. E., & Adachi, R. R. 1983. Bone resorption and mineral excretion in rats during spaceflight. *Am. J. Physiol.*, **244**, R327–31.

Caro, T. M. 1987. Indirect costs of play: Cheetah cubs reduce maternal hunting success. *Anim. Behav.*, **35**, 295–7.

Caro, T. M. 1994. *Cheetahs of the Serengeti Plains. Group Living in an Asocial Species*. Chicago: University Chicago Press.

Chivers, S. M., & Einon, D. F. 1982. Effects of early social experience on activity and object investigation in the ferret. *Dev. Psychobiol.*, **15**, 75–80.

Cloarec, A. 1982. Predatory success in the water stick insect: the role of visual and mechanical stimulations after moulting. *Anim. Behav.*, **30**, 549–56.

Close, R. I. 1972. Dynamic properties of mammalian skeletal muscles. *Physiol. Rev.*, **52**, 129–97.

Cooke, F., & McNally, C. M. 1975. Mate selection and colour preferences in lesser snow geese. *Behaviour*, **53**, 151–70.

Curtiss, S. 1977. *Genie: A Psycholinguistic Study of a Modern-day 'Wild Child'*. New York: Academic Press.

Einon, D. F., Morgan, M. J., & Kibbler, C. C. 1977. Brief periods of socialization and later behavior in the rat. *Develop. Psychobiol.*, **11**, 213–25.

Fagen, R. 1976. Exercise, play, and physical training in animals. In *Perspectives in Ethology, Vol. 2*. (eds P. P. G. Bateson & P. H. Klopfer), pp. 189–219. New York: Plenum.

Fagen, R. 1977. Selection for optimal age-dependent schedules of play behavior. *Am. Nat.*, **111**, 395–414.

Fagen, R. 1981. *Animal Play Behavior*. London and New York: Oxford University Press.

Fairbanks, L. A. 1993. Juvenile vervet monkeys: establishing relationships and practicing skills for the future. In *Juvenile Primates. Life History, Development, and Behavior*. (eds. M. E. Pereira & L. A. Fairbanks), pp. 211–27. New York: Oxford University Press.

Fillion, J. T., & Blass. M. E. 1986. Infantile experience with suckling odors determines adult sexual behavior in male rats. *Science*, **31**, 729–32.

Gans, C., & Gaunt, A. S. 1991. Muscle architecture in relation to function. *J. Biomech.*, **24**(suppl. 1), 53–65.

Gordon, T., & Pattullo, M. C. 1993. Plasticity of muscle fiber and motor type units. *Exercise Sport Sci. Rev.*, **21**, 331–62.

Govind, C. K. 1989. Asymmetry in lobster claws. *Am. Sci.*, **77**, 468–74.

Greenough, W. T., & Juraska, J. M. 1979. Experience-induced changes in brain fine structure: their behavioral implications. In *Development and Evolution of Brain Size: Behavioral Implications*. (eds. M. E. Hahn, Jensen C., & B. C. Dudek), pp. 295–320. New York: Academic Press.

Harcourt, R. 1991. Survivorship costs of play in the South American fur seal. *Anim. Behav.*, **42**, 509–11.

Hess, E. H. 1973. *Imprinting: Early Experience and the Developmental Psychobiology of Attachment*. New York: Van Nostrand.

Huck, U. W., & Banks, E. M. 1980. The effects of cross-fostering on the behavior of two species of North American lemmings, *Dicrostonyx groenlandicus* and *Lemmus trimucronatus*. III. Agonistic behavior. *Behaviour*, **73**, 261–76.

Immelmann, K. 1972. Sexual and other long-term aspects of imprinting in birds and other species. *Adv. Stud. Behav.*, **4**, 147–74.

Immelmann, K., & Suomi, S. 1981. Sensitive phases in development. In *Behavioral Development: The Bielefeld Interdisciplinary Project*. (eds. K.

Immelmann, G. W. Barlow, L. Petrinovich, & M. Main), pp. 395–431. Cambridge: Cambridge University Press.

Karban, R., & Myers, J. H. 1989. Induced plant responses to herbivory. *Ann. Rev. Ecol. Syst.*, **20**, 331–48.

Keeton, W. T. 1981. The ontogeny of bird orientation. In *Behavioral Development. The Bielefeld Interdisciplinary Project.* (eds. K. Immelmann, G. W. Barlow, L. Petrinovich, & M. Main), pp. 509–17. Cambridge: Cambridge University Press.

Kroodsma, D. E. 1981. Ontogeny of bird song. In *Behavioral Development. The Bielefeld Interdisciplinary Project.* (eds. K. Immelmann, G. W. Barlow, L. Petrinovich, & M. Main), pp. 518–32. Cambridge: Cambridge University Press.

Kuhl, P. K., Williams, K. A., Lacerda, F., Stevens, K. N., & Lindblom, B. 1992. Linguistic experience alters phonetic perception in infants by 6 months of age. *Science*, **255**, 606–8.

Larramendi, L. M. H. 1969. Analysis of synaptogenesis in the cerebellum of the mouse. In *Neurobiology of Cerebellar Evolution and Development.* (ed. R. R. Llinás), pp. 803–43. Chicago: American Medical Association.

Laughlin, M. H., Overholser, K. A. & Bhatte. M. J. 1989. Exercise training increases coronary transport reserve in miniature swine. *J. Appl. Physiol.*, **67**, 1140–9.

Locke, J. L. 1993. *The Child's Path to Spoken Language*. Cambridge, Massachusetts: Harvard University Press.

Lorenz, K. 1935. Der Kumpan in der Umwelt des Vogels. *J. Ornithol.*, **83**, 137–413.

Marler, P. 1970. A comparative approach to vocal learning: song development in white-crowned sparrows. *J. Comp. Physiol. Psychol. Monogr.*, **71**, 1–25.

Martin, P. 1984. The time and energy costs of play behaviour in the cat. *Z. Tierpsychol.*, **64**, 298–312.

Martin, P., & Caro, T. M. 1985. On the functions of play and its role in behavioral development. *Adv. Study Behav.*, **15**, 59–103.

Metcalfe, N. B., Huntingford, F. A. Graham, W. D., & Thorpe, J. E. 1989. Early social status and the development of life-history strategies in Atlantic salmon. *Proc. R. Soc. Lond., B*, **236**, 7–19.

Meyer, A. 1987. Phenotypic plasticity and heterochrony in *Cichlasoma managuense* (Pisces, Cichlidae) and their implications for speciation in cichlid fishes. *Evolution*, **41**, 1357–69.

Miller, M. N., & Byers, J. A. 1991. Energetic cost of locomotor play in pronghorn fawns. *Anim. Behav.*, **41**, 1007–13.

Morey, E. R., & Baylink, D. J. 1978. Inhibition of bone formation during space flight. *Science*, **201**, 1138–41.

Nieman, D. C. 1990. *Fitness and Sports Medicine: An Introduction*. Palo Alto, California: Bull Publishing.

Novakova, V., & Babicky, A. 1979. Role of early experience in social behavior of laboratory bred female rats. *Behav. Proc.*, **2**, 243–54.

Owens, N. W. 1975. Social play behaviour in free-living baboons, *Papio anubis*. *Anim. Behav.*, **23**, 387–408.

Palumbi, S. R. 1984. Tactics of acclimation: morphological changes of sponges in an unpredictable environment. *Science*, **225**, 1478–80.

Picarro, I. C., Barros Neto, T. L., Carrero de Teves, D., Silva, A. C., Denadai, D. S., Tarasantchi, J., & Russo, A. K. 1991. Effect of exercise during pregnancy, graded as a percentage of aerobic capacity: maternal and fetal responses of the rat. *Comp. Biochem. Physiol.*, **100A**, 795–99.

Rall, J. A. 1985. Energetic aspects of skeletal muscle contraction: implications of fiber types. *Exercise Sports Sc. Rev.*, **13**, 33–74.

Riegel, P. S. 1981. Athletic records and human endurance. *Am. Sci.*, **69**, 285–90.

Roy, R. R., Baldwin, K. M., & Edgerton, V. R. 1991. The plasticity of skeletal muscle: effects of neuromuscular activity. *Ex. Sports Sci. Rev.*, **19**, 269–312.

Siviy, S. M., & Atrens, D. M. 1992. The energetic costs of rough-and-tumble play in the juvenile rat. *Dev. Psychobiol.*, **25**, 137–48.

Sultan, E. S. 1987. Evolutionary implications of phenotypic plasticity in plants. *Evol. Biol.*, **21**, 127–78.

Thach, W. J. 1996. On the specific role of the cerebellum in motor learning and cognition: clues from PET activation and lesion studies in man. *Behav. Brain Sci.*, **19**, 411–31.

Thach, W. T., Goodkin, H. P., & Keating, J. G. 1992. The cerebellum and the adaptive coordination of movement. *Ann. Rev. Neurosci.*, **15**, 403–42.

Walker, C. B., & Byers, J. A. 1991. Heritability of locomotor play in house mice, *Mus domesticus*. *Anim Behav.*, **42**, 891–7.

Welles, R. E., & Welles, F. B. 1961. The bighorn of Death Valley. *U.S. Natl. Park Service. Fauna Ser. Monograph*, **6**, 242.

West, M. 1974. Social play in the domestic cat. *Am. Zool.* **14**, 427–36.

West-Eberhard, M. J. 1989. Phenotypic plasticity and the origins of diversity. *Ann. Rev. Ecol. Syst.*, **20**, 249–78.

Xia, Q. 1990. Morphological study of myocardial capillaries in endurance trained rats. *Br. J. Sports Med.*, **24**, 113–16.

11

Neurobiological substrates of play behavior: glimpses into the structure and function of mammalian playfulness

STEPHEN M. SIVIY
Department of Psychology, Gettysburg College, Gettysburg, PA 17325 USA

Introduction

There is no shortage of descriptive literature on play behavior, nor is there a lack of theories as to the putative functions and evolutionary origins of play. But despite this wealth of information and speculation, very little is known about how the brain is involved in playfulness. We can presume that those animals which play do so because their brains evolved in a way that favored the presence of certain types of neural circuitry, the activation of which results in behaviors that we can readily identify as being playful. Since it is now fairly well established that play is a distinct behavioral entity and not simply the juvenile form of more adult-typical behaviors (e.g., Fagen 1981), it can also be assumed that play would be represented by a distinct neural topography. This is not to say that overlap doesn't exist between brain areas involved in play and brain areas involved in other behaviors. For example, it would be expected that the pleasure derived from engaging in playful interactions taps into neural systems which code pleasure derived from other behaviors. Similarly, those systems which are involved in the actual patterning of movements exhibited during play would not be expected to differ greatly from those involved in motor patterning of other behaviors. The neural uniqueness of play may be in how these individual neural units are combined, and which types of stimuli initiate them.

Play has often been thought of as a distinctly mammalian behavior pattern (e.g., MacLean 1990). Such an assumption would make the search for neural systems involved in play relatively straightforward in that the search could be limited to parts of the brain unique to mammals. Indeed, MacLean (1985) proposed some time ago that play represented one of three 'signature' behaviors that separated mammals from reptiles, and that the appearance of play as a unique behavior probably coincided with the development of the limbic system. While the involvement of certain limbic structures in mammalian play is still very likely, the recent observation that turtles may exhibit behaviors looking remarkably like object play (Burghardt, Chapter 1), combined with earlier reports of avian play (listed in Fagen 1981), suggest that this behavior pattern may not be exclusively limbic in origin. Especially interesting about reports of non-mammalian play is the possibility that object play may have preceded social rough-and-tumble play. If this is the case and if object play in turtles is an evolutionary precursor to the more varied forms of play in mammals, then the evolutionary origins of playfulness may reside in pre-limbic regions of the brain.

One likely candidate for a neural representation of pre-limbic play would be the basal ganglia, as this area is very prominent in reptiles (MacLean 1990), and activity in this system could result in the motor patterns necessary to engage in object play. Indeed, when basal ganglia functioning is compromised in juvenile rats by neonatal 6–hydroxydopamine (6–OHDA) lesions, the motor patterning of play behavior is affected (Pellis et al. 1993). We could even envision the rough-and-tumble play of mammals as a higher-order form of object play in which the 'object' (i.e., the play partner) interacts with the player. This type of fast-paced social interaction, in which the intentions of the play partners must be constantly evaluated (e.g., Bekoff & Allen, Chapter 5; Pellis & Pellis 1996), might require more flexibility than the basal ganglia are capable of, hence an increased dependence upon limbic structures in mammalian play. While the search for a neural representation of play in mammals can continue unabated, recent reports of non-mammalian play raises several intriguing questions about the evolutionary origins of those neural systems responsible for mammalian play and opens up additional avenues of investigation that may not have otherwise been pursued.

In addition to lending insight into the evolutionary origins of play, studies that focus on the neural basis of play behavior can also be useful in generating testable hypotheses about possible functions of play. Any benefit which an animal gains from engaging in play behavior is likely to leave some evidence of that benefit in either neuronal structure (e.g., increased dendritic complexity or increased number of post-synaptic receptors) or in neuronal functioning (e.g., enhanced pre-synaptic release or increased sensitivity of receptors). Where to begin looking for this type of evidence could be indicated by the results of psychobiological studies designed to tease apart the relevant brain systems involved in mammalian playfulness.

Some methodological considerations

As is the case with any other behavior, our ability to study the neurobiological substrates of play is limited by the observational methods with which we can quantify the behavior, and the tools available to dissect and identify the relevant neural systems involved. The most commonly used tools in the psychobiological arsenal remain the lesion and pharmacological manipulations. While both of these techniques have their advantages and disadvantages, use of appropriate and multiple controls can minimize the disadvantages and lead to an increased understanding of the neural control of behavior. However, play behavior offers several challenges for those interested in pursuing an empirical analysis of the neurobiological substrates of this complex behavior. Play has traditionally been studied by ethologists, so it has most often been evaluated in the field over relatively long periods of time using traditional ethological techniques. Since many experimental manipulations are relatively acute, it has become necessary to adopt observational techniques which would allow for a satisfactory sample of the behavior to be observed in a limited amount of time. By restricting the opportunity to engage in play and focusing on a small number of highly reliable indicator variables to quantify levels of playfulness, this rich and complex behavior has become more amenable to tight experimental control (Panksepp et al. 1984). We now routinely use a paired-encounter design, in which rats are housed individually and allowed only brief (e.g., five minutes) daily opportunities to play in pairs, during which time the occurrence of certain specific motor patterns are quantified. The most common motor pattern observed during play is the pin. A pin is defined as occurring when one rat is on its dorsal surface with the other rat on top in what appears to be a 'dominant' posture. Pins are very easy to quantify during real-time observations, are correlated with other measures of play and show very high inter-rater reliability (Panksepp & Beatty 1980; Panksepp et al. 1984). As such, they represent a very good initial measure for determining overall levels of playfulness in a pair of rats. The addition of more sophisticated behavioral scoring techniques (e.g., Pellis 1988; Pellis & Pellis 1987, 1991) has provided an added level of descriptive power to the paired-encounter approach which allows for an even further dissection of individual components of rodent play. While my lab has begun to use some of Pellis' scoring techniques, most of the studies that I will be discussing in this chapter have used pins as the primary indicator variable for play.

Recent advances in molecular biology have also provided the psychobiologist with some very powerful tools which allow us to identify areas of the brain that may be especially relevant for the control of playfulness. By labelling the protein product of a particular class of immediate-early gene (e.g., *c-Fos*), we can visualize areas of the brain that have been most recently active (Dragunow & Faull 1989). This gives us a potential tool for visualizing those parts of the brain that are presumably most active during play behavior. Since this particular class of genes has also been implicated in processes associated with learning, memory and other forms of neural plasticity (e.g., Kaczmarek 1993; Morgan & Curran 1991), the use of this approach may also be useful in understanding the putative functions of play. We have begun to use this technique in my laboratory and the results of these studies will also be discussed.

We now have a fairly good arsenal of techniques with which to begin our search to identify relevant neural circuitry associated with mammalian playfulness. Whether such a circuit will be found, or whether bits and pieces of neural circuitry for play will be found embedded within the wiring pattern of systems associated with other behaviors, remains to be empirically determined. This chapter will focus primarily on what we currently know about some of the brain mechanisms that may be involved in the rough-and-tumble play of juvenile rats, although I hope that these data will apply to other mammals as well. While most of this discussion will focus on how the brain is involved in play, I will also attempt to speculate on how this knowledge can be used to better understand the evolutionary origins of play and the putative functions of play.

The neurochemistry of play

Much of my own research has focused on identifying neurochemical systems that are relevant for the occurrence of play in juvenile rats. The logic of this research is quite straightforward. Juvenile rats are administered (generally systemically) neuroactive compounds that are selective for a particular neurotransmitter system. These compounds are chosen for their ability to either enhance activity in that particular system, through any one of a number of mechanisms, or block activity in that system. Play can then be observed and quantified. Using this approach, a neurochemical picture of play in the rat is beginning to come into focus. Although this picture is far from complete, a number of neurotransmitter systems may be especially relevant for the regulation of play. Studies which have looked at the monoamines (dopamine, norepinephrine and serotonin) and brain opioids have resulted in the most intriguing data so far. These data suggest that dopamine may be important for invigorating the animal prior to a play bout, while norepinephrine (NE) and serotonin (5HT) may serve in a modulatory capacity once play is in full swing. Brain opioids also appear to have a modulatory role in regulating levels of play, perhaps by interacting with systems involved in pleasure–reward. Each of these systems will be discussed in turn.

Since a report by Beatty and colleagues (1982) describing the ability of psychomotor stimulants (e.g., amphetamine, methylphenidate) to reduce some aspects of rough-and-tumble play in juvenile rats, while increasing social investigation, there has been considerable interest in the extent to which any of these effects could be accounted for by the release of either dopamine or norepinephrine. However, a subsequent study failed to identify the underlying mechanism for amphetamine's effect on play (Beatty et al. 1984). Preliminary studies in my laboratory, using various antagonists for dopamine and NE, have also failed to identify the pharmacological nature of amphetamine's ability to affect play behavior. Since psychomotor stimulants have a wide range of neurochemical effects, traditional pharmacological blocking studies may not yield useful information concerning either the mechanism underlying amphetamine's effect on play or the endogenous control of the behavior. Use of more selective agonists and antagonists for the various monoaminergic systems, on the other hand, may be a more fruitful approach.

There is a substantial literature describing the involvement of dopamine in motivated behaviors, in processes associated with pleasure and reward, and in motor patterning (Blackburn et al. 1992; Le Moal & Simon 1991; Salamone 1994). Given that levels of rough-and-tumble play are under strong motivational control (Panksepp & Beatty 1980), and that play is both pleasurable (Calcagnetti & Schecter 1992; Humphreys & Einon 1981; Normansell & Panksepp 1990; Pellis & McKenna 1995) and motorically demanding, it could be predicted that dopamine is important in the control of play. Indeed, dopamine antagonists uniformly reduce play (Beatty et al. 1984; Holloway & Thor 1985; Niesink & VanRee 1989), while dopamine depleting 6–OHDA lesions impair the sequencing of behavioral elements observed during play (Pellis et al. 1993). On the other hand, studies which have examined the effects of stimulating dopamine receptors with agonists have yielded mixed results. Apomorphine has been reported to increase play within a narrow dose range (Beatty et al. 1984; Cox et al. 1984; Marshall et al. 1989), but this effect has not been very robust (e.g., Panksepp et al. 1987).

There is a strong trend for low doses of the more selective D_2 agonist quinpirole to increase play, but this has also been found to be not very reliable (Siviy et al. 1996). This pattern of findings is somewhat problematic in that one of the basic tenets of psychopharmacology research is that reciprocal behavioral effects ought to be obtained with agonists and antagonists for a particular neurochemical system. So using a straightforward psychopharmacological approach has not been that promising in illuminating the extent to which dopamine is involved in play.

Given the relative lack of success with standard pharmacological tools, we may have been asking the wrong questions about dopamine's involvement in play and, perhaps, not using the right tools. Since dopamine agonists and antagonists do not always yield reciprocal behavioral effects, dopamine may have a limited role in the actual display of the behavior. However, this does not necessarily mean that dopamine is unimportant for play to occur. One attempt to explain dopaminergic involvement in other motivated behaviors proposed that this particular transmitter system is important for the anticipatory component associated with motivated behavior (Blackburn et al. 1992; Salamone 1994). In particular, stimuli which have come to be associated with certain types of reinforcers (e.g., food, sexually receptive mate) are thought to increase activity in mesolimbic dopamine systems. Increased release of dopamine in mesolimbic terminal regions results in behaviors, such as increased locomotor activity and systematic searching, which increase the probability of coming into contact with the reinforcer. Two examples from personal experiences might be useful in better understanding this phenomenon. My family had a dog once named Ginger and, like most dogs, she loved to go on walks. After a while, Ginger began to associate certain stimuli with going for a walk and these stimuli had the effect of getting her extremely 'fired up' – barking, pacing, jumping up and down, running to and from the door. I have also noticed a similar phenomenon with my two year old daughter Kayla, who went through a phase (amazingly enough) during which she loved to have a bath. So, like Ginger who associated certain stimuli with a walk, Kayla came to associate certain stimuli with having a bath and when those stimuli were presented she would get up immediately from whatever she was doing, run over to the stairs and pace back and forth in front of the gate which blocked her way up the stairs. This became such a regular phenomenon that my older children would look forward to being the one allowed to mention the word 'bath' at the appropriate time and initiate this sequence of anticipatory behaviors. Both of these cases would be examples of a set of responses being made in anticipation of a rewarding situation and are presumably resulting from increased activity in mesolimbic dopamine pathways.

Can play also yield such an anticipatory response in rats? Juvenile rats will exhibit operant responses in order to play, in that rats will readily learn to traverse a maze when the opportunity to engage in a brief period of play is made the reward for choosing the correct path (Humphreys & Einon 1981; Normansell & Panksepp 1990). However, it is not clear from these data if this is a dopamine-mediated event. To address this issue, we designed a study to further quantify any anticipatory 'eagerness' that might be associated with play, while also assessing whether dopamine could be involved in the process. The design was a variation on the level-searching paradigm initially described by Mendelson & Pfaus (1989). In that study, adult male rats were placed in a bi-level chamber. Rats had free access to both levels and the number of times the rat changed levels was counted. After ten minutes in the apparatus, a receptive female was placed with the male and mating was allowed to occur. This procedure was repeated every third day for ten test days. The number of level changes increased steadily over the days of testing, presumably due to the male searching for the female. The increase in level-changing was described as an anticipatory component of sexual behavior and subsequent work found that this was accompanied by increased activity in mesolimbic dopamine pathways (Pfaus & Phillips 1991; Damsma et al. 1992).

In our design, two chambers that were connected by a horizontal tube were used (Figure 11.1). Individually housed juvenile rats were placed in the apparatus for five minutes and the number of crosses made through the tunnel from one chamber to the other was counted. At the end of the initial five minute period, the rats in the experimental condition were allowed to play with another juvenile rat for five minutes. Rats in the control group, on the other hand, were left in the apparatus for an additional five minutes with no opportunity to play. To control for total exposure to play, those rats in the control condition were given an opportunity to play in a different room approximately two to three hours after testing. This regimen was continued for eight days. As can be seen in Figure 11.2, rats that were given an opportunity to play showed significantly more tunnel crosses than the control rats as testing progressed, and the number of tunnel crosses exhibited by the experimental group was significantly reduced by a low dose of the dopamine antagonist haloperidol. This dose of haloperidol did not affect pinning when the rats were

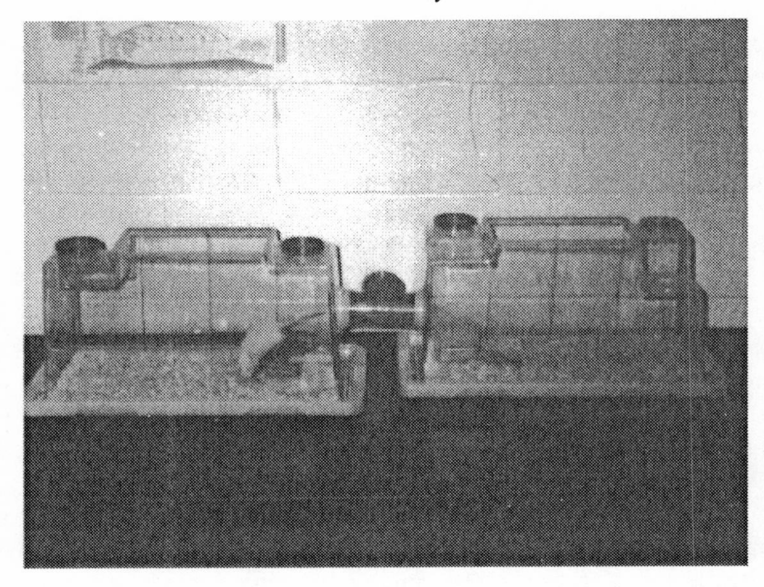

Figure 11.1. The apparatus which was used to quantify the anticipatory component associated with play behavior. The rat depicted in the apparatus is approximately 30 days old.

allowed to play with each other, suggesting that haloperidol was specifically blocking an anticipatory component to the behavior. While these data are still preliminary, they strongly suggest that stimuli associated with play elicit an anticipatory response in juvenile rats, and that this anticipatory response is likely to be dopamine mediated. If there is a parallel between what is being observed in our paradigm and what is being observed with other motivated behaviors, we can speculate further that this response is due to increased activity in mesolimbic dopamine systems.

Once a play bout has entered full stride, other neurotransmitters may begin to take on a more prominent role in modulating how the play bout will unfold. Two neurochemical systems that may have an especially important role in regulating levels of play are NE and 5HT. Much of the work that we have been doing with these two systems fits nicely into a model that was developed by Gaylord Ellison (1977). In Ellison's design, animals lived in a semi-naturalistic environment, with some sustaining noradrenergic lesions and some sustaining serotonergic lesions. The major focus of this work was to develop animal models of certain

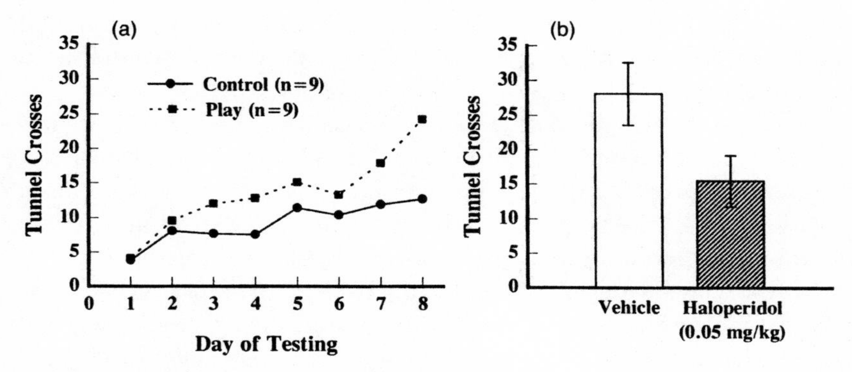

Figure 11.2. *A*: Mean number of tunnel crosses during a 5 minute test for those rats about to be given a 5 minute opportunity to play with a same-aged partner immediately following this test (Play) and for those not given an opportunity to play (Control). The number of tunnel crosses in the Play group increased significantly over days of testing, as evidenced by a significant Group X Day-of-testing interaction, $F(7, 112) = 2.62$, $p < 0.02$. *B*: Mean number of tunnel crosses for those rats in the Play group after either administration of the dopamine antagonist haloperidol or the vehicle (0.04M lactic acid). Injections were given subcutaneously 30 minutes before being placed in the testing apparatus. Rats were tested over two days, with the order of administration counterbalanced. The reduction in tunnel crosses after haloperidol was statistically significant, $t(8) = 4.52$, $p < 0.01$.

types of psychopathology, primarily depression and anxiety. Of particular interest within the present context, however, was the observation that animals with low levels of NE played less than the controls, while animals with low levels of 5HT played more. While the neurochemical selectivity of the lesions could not be ascertained from this particular report, these data suggest that NE and 5HT may have reciprocal effects on play, with NE facilitating and 5HT inhibiting playfulness. To further test this hypothesis, we have been investigating the effects of a wide variety of noradrenergic and serotonergic compounds on play.

Presynaptic release of both NE and 5HT is, to a large extent, under the modulatory control of pre-synaptic autoreceptors (Langer 1987). As molecules of transmitter are released, they will bind not only to post-synaptic sites, but also to receptors at pre-synaptic locations. This has the effect of keeping further release and/or synthesis of the transmitter in check. Since autoreceptors tend to be of a specific subtype (e.g., alpha-2 for norepinephrine and $5HT_{1A}$ for serotonin), compounds selective for these receptor subtypes can be exploited to study the potential role of these systems in play. Stimulating pre-synaptic autoreceptors will inhibit further release of transmitter, while blocking autoreceptors will result in continued release.

Clonidine is an alpha-2 noradrenergic receptor agonist which, at low doses, will inhibit the release of forebrain NE. When given to juvenile rats prior to a play bout, clonidine reduces play at very low doses (Normansell & Panksepp 1985a). Idazoxan and RX821002 are alpha-2 noradrenergic receptor antagonists which will block autoreceptors, thus preventing endogneous NE from binding to these receptors. This has the effect of increasing synaptic availability of NE (Dennis et al. 1987) and both of these compounds increase play in a dose-related manner (Siviy et al. 1990, 1994). It can not be stated with certainty that these behavioral effects are a result of these compounds acting at pre-synaptic sites. However, for the lack of a more compelling explanation for these data, we have been operating under the working hypothesis that this is the case and that alpha-2 antagonists increase play as a result of their ability to increase levels of NE in the forebrain. This has led to inquiries as to why higher levels of NE would lead to higher levels of play. Increased levels of NE increases the level of vigilance exhibited by an animal, which results in that animal being more attentive to environmental stimuli (Mason 1984; Oades 1985; Selden et al. 1990). While increases in vigilance might normally be in response to situations that are predictive of stressors (e.g., Levine et al. 1990), increased levels of NE in a non-stressful environment might be expected to have a different behavioral outcome. Since the rats are normally tested in the dark and in a chamber to which they are well-acclimated, increased vigilance might allow the animal to better direct its attention towards its play partner. In other words, there is no potential stressor in the immediate environment, so the animal is free to focus its attention on the most salient stimulus in its environment – another rat. If this is the case, then the introduction of disruptive, and potentially threatening, stimuli during a play bout should affect the ability of alpha-2 antagonists to increase play. In particular, the ability of an alpha-2 antagonist to increase play should decrease when disruptive stimuli are presented. We have tested this hypothesis by introducing various types of potentially disruptive stimuli, such as bright lights and noise, during a play bout. These stimuli were disruptive, since levels of play decreased when they were presented. But the effectiveness of RX821002 to increase play or clonidine to decrease play was not altered under these circumstances (Siviy, Baliko, Fleischhauer & Kerrigan, unpublished observations). So while alpha-2 agonists and antagonists have clear reciprocal effects on play, the underlying mechanism for these effects remains unclear.

Our understanding of the role that serotonin might have in the modulation of play has also been facilitated by the increased availability of compounds that act selectively at pre-synaptic sites. For example, 8–OH-DPAT is a selective agonist for the $5HT_{1A}$ receptor. This particular class of receptor is located at both pre- and post-synaptic sites, although many studies have shown that low doses of this compound tend to be somewhat selective for pre-synaptic autoreceptors (e.g., Carboni & DiChiara 1989; Hjorth et al. 1982). In some of our early work with this compound, we repeatedly observed a strong trend for low autoreceptor-selective doses to increase pinning. However, these increases were not always significant (reminiscent of our work with quinpirole). At the time of these tests, the only version of this compound that was commercially available was an enantiomeric mixture. When both the full agonist (R(+)-8–OH-DPAT) and partial agonist (S(-)-8–OH-DPAT) became separately available, we tested these in our paradigm and found that the full agonist significantly increased pinning at low doses and decreased pinning at higher doses. The partial agonist, on the other hand, failed to increase play at any dose and reduced pinning at higher doses (Siviy, Fleischhauer & Baliko, unpublished observations). We have replicated this finding and have also begun to analyze videotapes of these play bouts. While these data are still preliminary, it appears as if R(+)-8–OH-DPAT may increase pinning by increasing the likelihood that an animal will rotate completely to a supine position following a playful attack to its nape. Decreasing 5HT activity may then increase play by altering the responsiveness of the rats to playful overtures by conspecifics. Conversely, increasing synaptic availability of serotonin by either facilitating release with fenfluramine (Panksepp et al. 1987) or by blocking re-uptake with fluoxetine (Knutson et al. 1996; also unpublished observations from our lab.) reduces play, as does post-synaptic stimulation with the $5HT_2$ agonist quipazine (Normansell & Panksepp 1985b). These data regarding the 5HT system, combined with those looking at NE, tend to support Ellison's (1977) earlier observations and suggest that these two systems have reciprocal influences on playfulness.

Another neurochemical system that has received considerable attention is the endogenous opioids. The blockage of opioid receptors with antagonists selective for the μ receptor, such as naloxone, naltrexone or β-funaltrexamine, reliably reduces play (Beatty & Costello 1982; Niesink & Van Ree 1989; Panksepp et al. 1985; Siegel & Jensen 1986;

Vanderschuren et al. 1995a), while stimulation with μ receptor agonists such as morphine and fentanyl increases play (Niesink & Van Ree 1989; Panksepp et al. 1985; Vanderschuren et al. 1995a; 1995b). These data are consistent with other reports showing that play decreases the binding of [^{3}H]- diprenorphine while rats are playing (Panksepp & Bishop 1981; Vanderschuren et al. 1995c, 1995d), indicating higher circulating levels of the endogenous ligand. So it appears as if increased activity in opioid systems facilitates playfulness, although the mechanism for this enhancement is not clear. Panksepp and colleagues (e.g., Panksepp et al. 1984, 1985) have suggested that increased opioid activity may result in a state of 'social comfort' which would bias behavioral output towards playful interactions in juveniles, while others (Vanderschuren et al. 1995a, 1995b, 1995c, 1995d) have proposed that opioids enhance the rewards associated with playing.

From data such as those described above, we can see a neurochemical picture starting to develop. While the involvement of other neurochemical systems in the regulation of play is likely to unfold as more research is done, work on those systems mentioned above has yielded the most promising data so far. One possible scenario for this preliminary neurochemical story might go as follows. Stimuli which predict a playful experience would result in increased activity in the dopaminergic mesolimbic pathway. This would result in an increased release of dopamine in mesolimbic terminal areas, such as the prefrontal cortex and nucleus accumbens, resulting in energization of the animal and behavior patterns that would increase the probability of a playful interaction. Because of the diffuse nature of noradrenergic, serotonergic and opioid pathways, these systems are likely to exert a more modulatory influence on how the play bout will unfold. Increased noradrenergic activity may enhance the ability of a rat to focus its attention on the task at hand (i.e., playing), while increased opioid activity may enhance the pleasure associated with playing. For all of this to happen, serotonin levels must also be low, although how this involvement translates into specific aspects of the behavior cannot be gleaned from the present data.

Is there a distinct neural circuit for play?

Neurochemical data suggest little about the actual circuitry that might be involved in playfulness. When psychoactive compounds are administered systemically, there is little anatomical specificity associated with any effects, especially when working with systems as diffuse as NE, 5HT and opioids. While much can be gained by infusing chemicals directly into specific brain regions, the fragile nature of the juvenile skull makes this option a challenging task, although not an impossible one (e.g., Jalowiec et al. 1989). Given these limitations, different experimental approaches must be applied to address questions regarding specific neural circuitry. The most common approach for determining whether a particular brain structure is necessary for the presence of a particular behavior remains the lesion technique, and this has been used with some success in isolating neural areas especially relevant for play (Panksepp et al. 1984, 1995; Pellis et al. 1992, 1993; Siviy & Panksepp 1985, 1987). Functional imaging techniques could also provide valuable insight into those areas of the brain which are most active during play bouts. For example, positron emission tomography (PET), functional magnetic resonance imaging (fMRI) and high-resolution electroencephalographic (EEG) recording have been used with great success to visualize neural activity during specific cognitive tasks in humans (e.g., Posner & Raichle 1994). While these technniques may eventually be useful in imaging the human brain during playful activities, they are not very practical for visualizing neural activity in rats during the fury of rough and tumble play.

An alternative to these types of imaging techniques takes advantage of recent advances in molecular biology. Post-synaptic stimulation of neurons initiates a wide variety of responses, including the transcription of certain types of genes, known as immediate-early genes (e.g., c-*fos* and c-*jun*). As activity in a particular brain region increases, so does the transcription of these genes. By using standard immunohistochemical techniques to label the protein product of these genes, we can determine whether a particular area of the brain has been recently active (Dragunow & Faull 1989). Given the relative ease with which this technique can be applied, we hoped to use it as a means for identifying relevant brain areas involved in play. To begin our search, we evaluated the available lesion data to provide some direction. Rough-and-tumble play involves a considerable amount of somatosensory input and previous work has shown that regions of the brain which are involved in processing certain types of somatosensory information are particularly relevant for normal play to occur. For example, cells in the parafascicular area of the thalamus (PFA) receive somatosensory information of a very diffuse nature, having large receptive fields and high thresholds for excitation (e.g., Mountcastle 1984). This is in contrast to cells in the ventrobasal thalamic complex, which receive somatosensory information of a very detailed nature, have small receptive fields, low thresholds for excitation and provide a somatotopic map of the body surface (Mountcastle 1984). While lesions to the ventrobasal thalamic area have minimal effects on play (Siviy & Panksepp 1987), lesions to the PFA substantially reduce levels of pinning, while having minimal effects on other behaviors (Siviy & Panksepp 1985; 1987). It was suggested on the basis of these data that the PFA might be a major interface for integrating somatosensory information encountered during play and relaying this information in a manner that facilitates the motor patterns used for playing. If so, cells in this area would be expected to be very active during play and we should be able to quantify this activity by labeling for the presence of the c-Fos protein using standard immunohistochemical techniques.

To test this hypothesis, one group of juvenile rats was given a 30 minute opportunity to engage in rough-and-tumble play and was then returned individually to their home cages (these rats were all housed individually to maximize the amount of play exhibited during the 30 minute session). After 90 minutes in their home cage, the animals were sacrificed, and the brains removed, and processed for identification of the presence of the c-Fos protein. Rats in the control group were placed individually in the test chamber for 30 minutes, then placed back in their home cages for 90 minutes and had their brains processed in the same way as the 'play' rats. Of particular interest in this initial experiment was the number of Fos-positive neurons in the PFA. Significant increases in Fos levels were detected in the dorsal and medial portions of the thalamus (which included the PFA), for the rats in the play condition (67% over control levels). However, these increases were not limited to the PFA. Substantial increases in Fos levels were also observed in all areas of cortex that were analyzed (parietal, temporal, occipital), as well as the hippocampus (70–90% over control levels). These data show that rough-and-tumble play activates many areas of the brain. In retrospect, it seems hardly surprising that a good bit of the brain is active during a bout of play. Anyone who has observed two young animals in a vigorous bout of play would have no trouble accepting this. But can this information tell us anything about the neurobiological substrates of play?

The circumstances under which this particular experiment was conducted may not have been ideal for the purpose that we set out to accomplish. Animals that are freely engaging in uncontrolled bouts of play may be producing high levels of background noise, preventing any specific area from standing out. Perhaps future studies that use this technique could attempt to isolate different components of play in order to reduce the background noise. In any event, while these data did not provide the breakthrough that we were hoping for in identifying a neural map of play, they do point out limitations associated with this technique and may also provide a glimpse into one possible function of play, to be discussed in the next section.

Tentative hypotheses regarding the function of play

By studying the neural substrates of play, we hope to be in a better position for proposing reasonable hypotheses about the consequences associated with engaging in playful behaviors. In particular, this approach allows us to generate hypotheses regarding function that are at least consistent with what we know about the neural substrates involved in play. Based on the work that has been discussed in the present chapter, there seems to be enough evidence to entertain two tentative hypotheses regarding the functions of play. These are (1) that play may facilitate coping with certain types of environmental stressors, and (2) that play may provide a means to facilitate learning and creativity.

It seems clear from our work with the monoamines that dopamine, norepinephrine and serotonin are involved in the orchestration of various aspects of play behavior. Monoamines are also important for coordinating an organism's response to stress (Goldstein et al. 1996), so any behavior that involves these systems in as global a manner as play should be expected to alter the future sensitivity of these systems (e.g., Antelman et al. 1992). While the extent to which a *specific* social experience such as play may alter the future sensitivity of monoaminergic systems has not been directly tested, there is evidence that early social experience can alter the functional characteristics of these systems. For example, rats that have been reared in isolation have higher levels of both norepinephrine and dopamine than those reared in groups (Heritch et al. 1990; Nishikawa et al. 1976; Weinstock et al. 1978) and lower levels of serotonin (Jones et al. 1992), although others have reported no effect of isolation rearing on norepinephrine (Jones et al. 1991 1992). Isolation rearing also increases tyrosine hydroxylase activity (Toru 1982), enhances transcription of tyrosine hydroxylase mRNA (Angulo et al. 1991) and down-regulates dopamine D2 receptors (Hall et al. 1991). Monoaminergic functioning may also be affected by early social experiences, as isolation-reared rats are more sensitive to the effects of amphetamine (Jones et al. 1990, 1992).

These data suggest that social experiences during the age when play is most often observed can have an impact on the structure and function of those neurochemical systems which appear to be involved in play behavior. These data do not, however, address the extent to which the effects observed with isolation rearing are due to a lack of play, nor do they address whether there is a connection between early social experience, monoaminergic systems and later responsiveness to stress (but see Einon et al. 1978). However, isolation rearing has also been reported to impair an animal's ability to deal effectively with social stressors (e.g., dealing with an aggressor), while having no effect on responsiveness to non-social stressors, such as an uncontrollable shock (Hol et al. 1994). If any of the above-mentioned effects are specifically due to a lack of play, then play experience may result in a brain that is better able to deal effectively with specific types of social stressors. In other words, those who have had an ample opportunity to engage in play as juveniles may be better equipped at a neural level to 'roll with the punches' associated with daily social interactions than those who haven't had this opportunity. Whether this is the case needs be more vigorously pursued.

Another hypothesis which is consistent with the data described above is that play may facilitate learning and creativity. This is a hypothesis which has its origins in the writings of Jean Piaget (e.g., Piaget 1964), and which is very consistent with the preliminary data we collected on play-induced increases in Fos levels. For example, the one consistent finding that we observed when analyzing the Fos patterns of rats that have recently played was a widespread activation of the cortex and certain subcortical structures (e.g., thalamus and hippocampus). While this global increase in neural activity can be viewed as problematic for identifying critical neural areas involved in play behavior, it may actually shed considerable light on an advantage that the playing brain may have over the non-playing brain. Learning is activity-dependent, so a brain that is very active is in a much better position to learn than a brain that is less active. In fact, the transcription of immediate-early genes such as c-*fos* and c-*jun* is believed to be a critical link in the molecular machinery associated with learning and memory (e.g., Kaczmarek 1993; Morgan & Curran 1991). Widespread activation of the brain would also tend to facilitate the formation of novel connections between brain areas that might not normally be connected, perhaps enhancing creativity. Predictions that follow from this hypothesis are that learning should be more efficient within the temporal boundaries of play bouts and creativity should be enhanced by playful experiences. While the former prediction has yet to be adequately tested, there is some support for the latter prediction. Dansky & Silverman (1973) reported that pre-schoolers who were allowed to freely play with common objects listed more novel uses for those objects than children who either were told what to do with the objects or were allowed to do some unrelated activity. These investigators concluded that the opportunity to play allowed these children to be more creative in their possible uses for these objects. While this experiment demonstrates a short-term enhancement of creativity, it isn't clear whether such a benefit stays with the child beyond the temporal boundaries of the play experience.

Conclusions

Not long ago, I was with my two year old daughter while she was playing with some of her toys. As I watched her intently manipulating some blocks and toy people, her face struck me as one that was beginning to show signs of some truly complex cognitive processes (or at least complex for a two year old). Play could be the first opportunity that the young of many mammalian and avian species have to engage in those mental gymnastics that we call cognition (e.g., Piaget 1964). Prior to the age when play begins, very little complex thought is needed. But when the young child or animal begins to 'venture forth into the world' and starts to interact in a more systematic and purposeful way with the environment, the possibilities for what might be encountered and how these encounters may turn out become vast. So it might be that species which can afford the luxury of expending a little extra energy (e.g., Burghardt 1988; Martin 1984; Siviy & Atrens 1992) will gravitate towards testing the various possibilities that they are confronted with as soon as they are able to do so.

One question that arises from this somewhat intuitive observation is whether it can be reconciled with the two hypotheses mentioned above. If we think of cognition as a mental representation of how one might interact with environmental stimuli (whether the stimulus is some object or another member of their species), then play could facilitate the development of this process by resulting in a brain that has a greater number of response options. By recruiting activity over many neural areas during play, mental representations of playful encounters could be more easily placed in a cognitive lexicon for retrieval at a later time. By tapping into neural systems involved in adaptation to stress (e.g., monoamines), this brain could also be better at coping with those instances when the behavior that arises from the cognition results in some undesirable consequence (e.g., social defeat). Thus, play could improve the animal's ability to manage environmental complexity in the future (see also Einon et al. 1978 and Enomoto 1990). The ultimate value associated with any of these hypotheses, whether they be intuitively or empirically derived, will be determined by the extent to which they can be empirically scruitinized and tested in the laboratory and field.

References

Angulo, J. A., Printz, D., Ledoux, M., & McEwen, B. S. 1991. Isolation stress increases tyrosine hydroxylase mRNA in the locus coeruleus and midbrain and decreases proenkephalin mRNA in the striatum and nucleus accumbens. *Mol. Brain Res.,* **11**, 301–8.

Antelman, S. M., Kocan, D., Knopf, S., Edwards, D. J., & Caggiula, A. R. 1992. One brief exposure to a psychological stressor induces long-lasting time-dependent sensitization of both the cataleptic and neurochemical responses to haloperidol. *Life Sci.*, **51**, 261–6.

Beatty, W. W., & Costello, K. B. 1982. Naloxone and play fighting in juvenile rats. *Pharmacol. Biochem. Behav.*, **17**, 905–7.

Beatty, W. W., Dodge, A. M., Dodge, L. J., White, K., & Panksepp, J. 1982. Psychomotor stimulants, social deprivation and play in juvenile rats. *Pharmacol. Biochem. Behav.*, **16**, 417–22.

Beatty, W. W., Costello, K. B., & Berry, S. L. 1984. Suppression of play fighting by amphetamine: Effects of catecholamine antagonists, agonists and synthesis inhibitors. *Pharmacol. Biochem. Behav.*, **20**, 747–55.

Blackburn, J. R., Pfaus, J. G., & Phillips, A. G. 1992. Dopamine functions in appetitive and defensive behaviours. *Prog. Neurobiol.*, **39**, 247–9.

Burghardt, G. M. 1988. Precocity, play, and the ectotherm-endotherm transition: Profound reorganization or superficial adaptation? In: *Handbook of Behavioral Neurobiology, Volume 9, Developmental Psychobiology and Behavioral Ecology* (ed. E. M. Blass), pp. 107–48, New York: Plenum Press.

Calcagnetti, D. J., & Schechter, M. D. 1992. Place conditioning reveals the rewarding aspect of social interaction in juvenile rats. *Physiol. Behav.*, **51**, 667–72.

Carboni, E., & DiChiara, G. 1989. Serotonin release estimated by transcortical dialysis in freely-moving rats. *Neuroscience,* **32**, 637–45.

Cox, J. F., Schoen, L., Normansell, L., Rossi III, J., Siviy, S., & Panksepp, J. 1984. Dopaminergic substrates of play. *Soc. Neurosci. Abst.*, **10**, 1177.

Damsma, G., Pfaus, J. G., Wenkstern, D., Phillips, A. G., & Fibiger, H. C. 1992. Sexual behavior increases dopamine transmission in the nucleus accumbens and striatum of male rats: Comparison with novelty and locomotion. *Behav. Neurosci.*, **106**, 181–91.

Dansky, J. L., & Silverman, I. W. 1973. Effects of play on associative fluency in preschool aged children. *Dev. Psychol.*, **9**, 38–44.

Dennis, T., L'Heureux, R., Carter, C., & Scatton, B. 1987. Presynaptic alpha-2 adrenoceptors play a major role in the effects of idazoxan on cortical noradrenaline release (as measured by in vivo dialysis) in the rat. *J. Pharmacol. Exp. Ther.* **241**, 642–9.

Dragunow, M., & Faull, R. 1989. The use of c-*fos* as a metabolic marker in neuronal pathway tracing. *J. Neurosci. Meth.*, **29**, 261–5.
Einon, D., Morgan, M., & Kibbler, C. 1978. Brief periods of socialization and later behavior in the rat. *Dev. Psychobiol.*, **11**, 213–25.
Ellison, G. D. 1977. Animal models of psychopathology: The low-norepinephrine and low-serotonin rat. *Am. Psychologist*, **32**, 1036–45.
Enomoto, T. 1990. Social play and sexual behavior of the Bonobo (*Pan paniscus*) with special reference to flexibility. *Primates*, **31**, 469–80.
Fagen, R. 1981. *Animal Play Behavior*. New York: Oxford University Press.
Goldstein, L. E., Rasmusson, A. M., Bunney, B. S., & Roth, R. H. 1996. Role of the amygdala in the coordination of behavioral, neuroendocrine, and prefrontal cortical monoamine responses to psychological stress in the rat. *J. Neurosci.*, **16**, 4787–98.
Hall, F. S., Wilkinson, L. S., Kendall, D. A., Marsden, C. A., & Robbins, T. W. 1991. Effects of isolation rearing on indices of dopamine function in the rat. *Soc. Neurosci. Abst.*, **17**, 731.
Heritch, A. J., Henderson, K., & Westfall, T. C. 1990. Effects of social isolation on brain catecholamines and forced swimming in rats: Prevention by antidepressant treatment. *J. Psych. Res.*, **24**, 251–8.
Hjorth, S., Carlsson, A., Lindberg, P., Sanchez, D., Wikstrom, H., Arvidsson, L.-E., Hacksell, U., & Nilsson, J. L. G. 1982. 8–Hydroxy-2–(Di-n-propylamino)tetralin, 8–OH-DPAT, a potent and selective simplified ergot congener with central 5–HT-receptor stimulating activity. *J. Neural Transmission*, **55**, 169–88.
Hol, T., Koolhaas, J. M., & Spruijt, B. M. 1994. Consequences of short term isolation after weaning on later adult behavioural and neuroendocrine reaction to social stress. *Behav. Pharmacol.*, **5**, 88–9.
Holloway, W. R., & Thor, D. H. 1985. Interactive effecs of caffeine, 2–chloroadenosine and haloperidol on activity, social investigation and play fighting of juvenile rats. *Pharmacol. Biochem. Behav.*, **22**, 421–6.
Humphreys, A. P., & Einon, D. R. 1981. Play as a reinforcer for maze-learning in juvenile rats. *Anim. Behav.*, **29**, 259–70.
Jalowiec, J. E., Calcagnetti, D. J., & Fanselow, M. S. 1989. Suppression of juvenile social behavior requires antagonism of central opioid systems. *Pharmacol. Biochem. Behav.*, **33**, 697–700.
Jones, G. H., Marsden, C. A., & Robbins, T. W. 1990. Increased sensitivity to amphetamine and reward-related stimuli following social isolation in rats: Possible disruption of dopamine-dependent mechanisms of the nucleus accumbens. *Psychopharmacology*, **102**, 364–72.
Jones, G. H., Marsden, C. A., & Robbins, T. W. 1991. Behavioural rigidity and rule-learning deficits following isolation-rearing in the rat: Neurochemical correlates. *Behav. Brain Res.*, **43**, 35–50.
Jones, G. H., Hernandez, T. D., Kendall, D. A., Marsden, C. A., & Robbins, T. W. 1992. Dopaminergic and serotonergic function following isolation rearing in rats: Study of behavioural responses and postmortem in vivo neurochemistry. *Pharmacol. Biochem. Behav.*, **43**, 17–35.
Kaczmarek, L. 1993. Molecular biology of vertebrate learning: Is c-fos a new beginning *J. Neurosci. Res.*, **34**, 377–81.
Knutson, B., Panksepp, J., & Pruitt, D. 1996. Effects of fluoxetine on play dominance in juvenile rats. *Aggressive Behav.*, **22**, 297–307.

Langer, S.Z. 1987. Presynaptic regulation of monoaminergic neurons. In: *Psychopharmacology: The Third Generation of Progress* (ed. H. Y. Meltzer), pp. 151–7, New York: Raven Press.

Le Moal, M., & Simon, H. 1991. Mesocorticolimbic dopaminergic network: Functional and regulatory roles. *Physiol. Rev.*, **71**, 155–234.

Levine, E. S., Litto, W. J., & Jacobs, B. L. 1990. Activity of cat locus coeruleus noradrenergic neurons during the defense reaction. *Brain Res.*, **531**, 189–95.

MacLean, P. D. 1985. Brain evolution relating to family, play, and the separation call. *Arch. Gen. Psychiatr.*, **42**, 405–17.

MacLean, P. D. 1990. *The Triune Brain in Evolution: Role in Paleocerebral Functions*. New York: Plenum Press.

Marshall, H. M., Pellis, S. M., Pellis, V. C., & Teitelbaum, P. 1989. Dopaminergic drugs differentially affect attack versus defense in play fighting by juvenile rats. *Soc. Neurosci. Abst.*, **15**, 1157.

Martin, P. 1984. The time and energy costs of play behaviour in the cat. *Zeit. Tierpsychol.*, **64**, 298–312.

Mason, S. T. 1984. *Catecholamines and Behaviour*. Cambridge: Cambridge University Press.

Mendelson, S. D., & Pfaus, J. G. 1989. Level searching: A new assay of sexual motivation in the male rat. *Physiol. Behav.*, **45**, 337–41.

Morgan, J. I., & Curran, T. 1991. Stimulus transcription coupling in the nervous system: Involvement of the inducible proto-oncogenes fos and jun. *Ann. Rev. Neurosci.*, **14**, 421–51.

Mountcastle, V. B. 1984. Central nervous mechanisms in mechanoreceptive sensibility. In: *Handbook of Physiology,* Section I, Volume II. (ed. I. Darian-Smith), Bethesda: American Physiological Society.

Niesink, R. J. M., & Van Ree, J. M. 1989. Involvement of opioid and dopaminergic systems in isolation-induced pinning and social grooming of young rats. *Neuropharmacology*, **28**, 411–18.

Nishikawa, T., Kajiwara, Y., Kono, Y., Sono, T., Nagasaki, N., & Tanaka, M. 1976. Different effects of social isolation on the levels of brain monoamines in post-weaning and young-adult ras. *Folia Psychiatr.Neurol. Japonica.*, **30**, 57–63.

Normansell, L. A., & Panksepp, J. 1985a. Effects of clonidine and yohimbine on the social play of juvenile rats. *Pharmacol. Biochem. Behav.*, **22**, 881–3.

Normansell, L., & Panksepp, J. 1985b. Effects of quipazine and methysergide on play in juvenile rats. *Pharmacol. Biochem. Behav.*, **22**, 885–7.

Normansell, L., & Panksepp, J. 1990. Effects of morphine and naloxone on play-rewarded spatial discrimination in juvenile rats. *Dev. Psychobiol.*, **23**, 75–83.

Oades, R. D. 1985. The role of noradrenaline in tuning and dopamine in switching between signals in the CNS. *Neurosci. Biobehav. Rev.*, **9**, 261–82.

Panksepp, J., & Beatty, W. W. 1980. Social deprivation and play in rats. *Behav. Neural Biol.*, **30**, 197–206.

Panksepp, J., & Bishop, P. 1981. An autoradiographic map of [^{3}H]diprenorphine binding in rat brain: effects of social interaction. *Brain Res. Bull.*, **7**, 405–10.

Panksepp, J., Jalowiec, J., DeEskinazi, F. G., & Bishop, P. 1985. Opiates and play dominance in juvenile rats. *Behav. Neurosci.*, **99**, 441–53.

Panksepp, J., Siviy, S. M., & Normansell, L. 1984. The psychobiology of play: Theoretical and methodological considerations. *Neurosci. Biobehav. Rev.*, **8**, 465–92.

Panksepp, J., Normansell, L., Cox, J. F., Crepeau, L. J., & Sacks, D. S. 1987. Psychopharmacology of social play. In: *Ethopharmacology of agonistic behaviour in animals and humans* (eds. B. Olivier, J. Mos, & P. F. Brain), pp. 132–41, Dordrecht, Holland: Martinus Nijhoff Publishers.

Pellis, S. M. 1988. Agonistic versus amicable targets of attack and defense: Consequences for the origin, function, and descriptive classification of play-fighting. *Aggressive Behav.*, **14**, 85–104.

Pellis, S. M., & McKenna, M. 1995. What do rats find rewarding in play fighting? An analysis using drug-induced non-playful partners. *Behav. Brain Res.*, **68**, 65–73.

Pellis, S. M., & Pellis, V. M. 1987. Play-fighting differs from serious fighting in both target of attack and tactics of fighting in the laboratory rat *Rattus norvegicus. Aggressive Behav.*, **13**, 227–52.

Pellis, S. M., & Pellis, V. C. 1991. Attack and defense during play fighting appear to be motivationally independent behaviors in muroid rodents. *Psychol. Record,* **41**, 175–84.

Pellis, S. M., & Pellis, V. C. 1996. On knowing it's only play: The role of play signals in play fighting. *Aggression Violent Behav.*, **1**, 249–68.

Pellis, S. M., Pellis, V. C., & Whishaw, I. Q. 1992. The role of the cortex in play fighting by rats: Developmental and evolutionary implications. *Brain Behav. Evol.*, **39**, 270–84.

Pellis, S. M., Casteneda, E., McKenna, M. M., Tran-Nguyen, L. T., & Whishaw, I. Q. 1993. The role of the striatum in organizing sequences of play fighting in neonatally dopamine-depleted rats. *Neurosci. Lett.*, **158**, 13–15.

Pfaus, J. G., & Phillips, A. G. 1991. Role of dopamine in anticipatory and consummatory aspects of sexual behavior in the male rat. *Behav. Neurosci.*, **105**, 727–43.

Piaget, J. 1964. *Play, Dreams and Imitation in Childhood.* New York: W.W. Norton and Company.

Posner, M. I., & Raichle, M. F. 1994. *Images of Mind.* New York: Scientific American.

Salamone, J. D. 1994. The involvement of nucleus accumbens dopamine in appetitive and aversive motivation. *Behav. Brain Res.*, **61**, 117–33.

Selden, N. R. W., Robbins, T. W., & Everitt, B. J. 1990. Enhanced behavioral conditioning to context and impaired behavioral and neuroendocrine responses to conditioned stimuli following cerueocortical noradrenergic lesions: Support for an attentional hypothesis of central noradrenergic function. *J. Neurosci.*, **10**, 531–39.

Siegel, M. A., & Jensen, R. A. 1986. The effects of naloxone and cage size on social play and activity in isolated young rats. *Behav. Neural Biol.*, **45**, 155–68.

Siviy, S. M., & Atrens, D. M. 1992. The energetic costs of rough-and-tumble play in the juvenile rat. *Dev. Psychobiol.*, **25**, 137–48.

Siviy, S. M., & Panksepp, J. 1985. Dorsomedial diencephalic involvement in the juvenile play of rats. *Behav. Neurosci.*, **99**, 1103–13.

Siviy, S. M., & Panksepp, J. 1987. Juvenile play in the rat: Thalamic and brain stem involvement. *Physiol. Behav.*, **41**, 103–14.

Siviy, S. M., Atrens, D. M., & Menendez, J. A. 1990. Idazoxan increases rough-and-tumble play, activity and exploration in juvenile rats. *Psychopharmacology*, **100**, 119–23.
Siviy, S. M., Fleischhauer, A. E., Kuhlman, S. J., & Atrens, D. M. 1994. Effects of alpha-2 adrenoceptor antagonists on rough-and-tumble play in juvenile rats: evidence for a site of action independent of non-adrenoceptor imidazoline binding sites. *Psychopharmacology*, **113**, 493–9.
Siviy, S. M., Fleischhauer, A. E., Kerrigan, L. A., & Kuhlman, S. J. 1996. D_2 dopamine receptor involvement in the rough-and-tumble play behavior of juvenile rats. *Behav. Neurosci.*, **110**, 1–9.
Toru, M. 1982. Increased tyrosine hydroxylase activity in frontal cortex of rats after long-term isolation stress. *L'Encephale*, **8**, 315–17.
Vanderschuren, L. J. M. J., Niesink, R. J. M., Spruijt, B. M., & Van Ree, J. M. 1995a. Effects of morphine on different aspects of social play in juvenile rats. *Psychopharmacology*, **117**, 225–31.
Vanderschuren, L. J. M. J., Niesink, R. J. M., Spruijt, B. M., & Van Ree, J. M. 1995b. μ- and k-opioid receptor-mediated opioid effects on social play in juvenile rats. *Eur. J. Pharmacol.*, **276**, 257–66.
Vanderschuren, L. J. M. J., Stein, E. A., Wiegant, V. M., & Van Ree, J. M. 1995c. Social play alters regional brain opioid receptor binding in juvenile rats. *Brain Res.*, **680**, 148–56.
Vanderschuren, L. J. M. J., Stein, E. A., Wiegant, V. M., & Van Ree, J. M. 1995d. Social isolation and social interaction alter regional brain opioid receptor binding in rats. *Eur. Neuropsychopharmacol.*, **5**, 119–27.
Weinstock, M., Speizer, Z., & Ashkenazi, R. 1978. Changes in brain catecholamine turnover and receptor sensitivity induced by social deprivation in rats. *Psychopharmacology*, **56**, 205–9.

12

Play as an organizing principle: clinical evidence and personal observations

STUART BROWN
225 Crossroads Blvd. Box 341, Carmel, CA 93923 USA

Introduction

By taking the reader along the play-path of my personal and professional life, I encourage readers to examine their own play experiences, attitudes, and observations. By reviewing my experiences sequentially, a gradually evolving broad view of the importance of, and speculations about what play is and what playfulness does, emerges. Play is seen as a broad category of behavior, as basic in its phenomenology to smart complex animals as sleep and dreams, and as scientifically enigmatic. Its healthy presence seems necessary for the maintenance of flexibility and adaptability.

I am a physician-psychiatrist by training and practice and more recently have engaged in independent scholarship, educational film production and popular writing (Brown 1987, 1988, 1995; Brown & Cousineau 1990; Brown & Moses 1995). As a method useful to me in making better sense of the world and its parts, I have generally relied on clinical observation, first to demonstrate a phenomenon and then have gone searching for 'explanations' which best explain and characterize it. This has worked well for me as a physician, and has enriched explorations in other areas. The general subject of *play* in animals and humans has gradually emerged as a broad category of behavior which warrants fresh exploration (Brownlee 1997). Thus the views given here about this complex and slippery subject will reflect my current efforts to place it in context as I have encountered it. My goals for this article are that it will offer the play enthusiast and student fresh ways of viewing the subject and stimulate personal examination of the cultures and biology of playfulness. I look upon this effort as if I was engaged in telling my story of play.

As children, you and I and all safe and well-fed kids *experienced* play as important. From the moment of an infants earliest post feeding nipple play, it bubbles forth naturally and is the engine that drives much of the spontaneous activity of our childhood. It energizes children worldwide, fosters a child's play culture which differs in form and language from that of adults (Sutton-Smith 1974, 1972).

The play of my childhood was quite open. I enjoyed roughhousing with my peer group, non-organized school games, afternoon unsupervised play, exploratory roaming without much adult on-site oversight. The internal narratives which defined my childhood sense of reality were not atypical, and were similar to those depicted in some of the writings of Sutton-Smith (McMahon & Sutton-Smith 1995; Sutton-Smith et al. 1995). My recall of them has been enhanced by detailed clinical reviews of play with a myriad of patients, and by other opportunities that have jogged my memory as I have studied play. Then and now, these early private stream-of-consciousness childhood stories, inculcated by play, whether left to ones own inner life, or secretly shared with peers, are considered outrageous by the prevailing adult culture (Sutton-Smith 1981). Though culturally sculpted, the spontaneous narratives and stories of children appear structurally similar now to those of thirty or forty years ago (Sutton-Smith, personal communication).

I also was fortunate enough to have parents who interfered little in controlling the spontaneity of play, and had an extended family who freely permitted both childhood and adult play, a condition not unusual in the 1940s. But play was, in these contexts, unquestioned and never viewed analytically. It was, however, tacitly honored as important, except in pious moments of temporary throwbacks to Puritanical cultural over-control when the chaos and freedom of open play created adult discomfort and could not be welcomed by my conventional and religiously devout parents. The point of generalizing about these personal childhood experiences here is that most adults have significant amnesia regarding their own play unless systematically summoned to recall it, and they tend to become threatened by the very behavior they once spontaneously enjoyed (Sutton-Smith 1993; Pellegrini 1989).

Most adults also tend to compartmentalize their adult lives into work–play dichotomies, which is not the way of children. For children, virtually all of their non-survival activities are play. This admittedly sparse description of childhood play is not meant to be an ideological pitch for an unrealistic impractical permissive societal approach to play management, but rather should serve as a partial description to spark the reader to open the door of memory to his or her own play natural history and its surrounding cultures. This process often positively shifts adults to re-evaluate their attitudes towards play in their own and their children's lives.

I do not recall any specific transitional moments when duty, responsibility, productivity, and other 'adult' behaviors became more rewarded than play, or exactly why or when a work–play separation occurred, but it did, and seemed normal. Joseph Meeker examines this process in his book *The Comedy of Survival* (1997).

My first dawning awareness that play warranted clinical explanation came as a medical student rotating through the hospital pediatric services observing and helping to care for desperately ill infants and children. I became impressed that as sick children began to recover, often the first signal of their return to wellness was their erupting sense of humor, or other signs of a return of spontaneous play. Their playful ways were frequently the most reliable signs of impending recovery and often *preceded* positive changes in temperature levels or laboratory findings. Later experiences as a Navy family physician with a large pediatric practice reinforced these earlier observations but still did not place play in a particularly separate category nor lead me to see it as a subject worth investigating. Now, hospital pediatric departments regularly provide play areas for inpatients, recognizing its participation in the healing process.

After practicing general medicine in the Navy, I became a resident specializing first in internal medicine, and later completed formal training and became Board certified in psychiatry. I began to acquire more analytic and research skills, and started my academic career in clinical research and teaching at Baylor College of Medicine, Houston, Texas.

When a Rubella (viral German measles) epidemic hit Houston, the Texas Medical Center mobilized me to be of assistance to the victims, and, with David Freedman as the principal investigator, we began to follow the development and course of the disease in many seriously ill congenitally infected babies (focusing in-depth research on one in particular). As they slowly recovered, we used the then state of the art technology to assay the emergence of audition, sight, pain perception and other developing modalities which had been damaged by the viral assault (Brown & Freedman 1970).

Most of these infants were profoundly affected by their disease, having central nervous system damage as well as suffering peripheral defects. Many were clinically deaf and blind, often incapable of vocalizations or even withdrawal responses to pain. As the virus was cleared by their developing immune systems, and in the presence of loving and concerned parents, those who did not suffer profound irreversible damage slowly began to integrate their damaged brains. This 'slow motion' recovery allowed windows of observation normally obscured by already completed fetal maturation, or missed due to the normal avalanche of developmental changes during the first months of extrauterine life. By using the techniques of EEG evoked cortical potentials (modified by a computer of average transients), we were able to recognize the first signals indicating cortical integration of sight, sound or pain, seen before the babies showed obvious clear external responses to these sensory inputs. Recognizable repetitive EEG wave forms stimulated by specific visual objects, sounds or tactile stimuli, heralded the cortical capacity for recognition of these recognizable sensory signals which had been nonexistent while the virus was active, In uninfected newborns, definite visual, auditory and pain responses are generally already intact and are progressively and rapidly changing as maturation occurs. During this special time of observed recovery in the rubella children, I incidentally noted that *play behavior seemed to precede the clear EEG establishment of perceptual awareness.*

Play, as it emerged in this late 1960s research setting, was also a positive prognostic sign, and seemed to differentiate those whose nervous systems would integrate and heal, from those who were so permanently damaged they could not. I do not know the significance of this observation, but play was a sign of impending cognitive and perceptual normalcy in healing children. The notion that play behavior itself (viewed in 'slow motion') may have a sculpting action on neural patterning warrants further research exploration, and seems to place behavior occurring in advance of permanent neurologic changes.

During the time that the rubella studies were underway, I was appointed to a research team assigned to understand the motives and life of the Texas tower mass murderer Charles Whitman, and following that became the principal investigator of other clinical research projects that added new dimensions to my growing story of play. Whitman, a 25 year old architectural engineering student at The University of Texas, Austin, had, after killing his wife and mother, mounted the campus tower, and with deadly accurate fire, killed 17 and wounded 31, before being gunned down by vigilante and police crossfire. Whitman had been a model student, a supposed loving husband and son, at the age of 12 the USA's youngest Eagle scout, a Naval Reserved Officer Training Corp scholarship recipient, and an ex-Marine. The then Texas Governor John

Connally, who had himself been wounded in the John F. Kennedy assassination was urgent in his insistence that we discern what made Whitman tick, and the Texas legislature fully funded a task force to find answers. It was the late 1960s, and fear that assassins and havoc were taking over society reigned.

That in-depth study allowed us to dissect intricately Whitman's psychobiology. To accomplish this, we called in a broadly based, distinguished multidisciplinary group of consulting experts, ranging from pathologists and toxicologists, neuroanatomists and neurologists, to graphologists, sociologists, neuropsychologists, and other specialists. We conducted field interviews that reconstructed his life in as much detail as possible, and catalogued every aspect of his behavior available. Our investigation included such things as the written and verbal recollections of the family doctor who delivered him and cared for his mother, (for example, we learned she was kicked in the stomach by Whitman's father when 6 months pregnant with Charles). We conducted extensive interviews with his large extended family, and examined childhood drawings and nursery school teacher anecdotes. We systematically organized additional lore from friends, carefully read his lengthy diaries, accessed his medical and military records, and reviewed home movies, all school records, and snapshots. As we recreated his life we began to gain a consistent recognition of his personal ecology and development, weighing as many factors that defined and predicted his behavior as we could. The full compilation of data was compelling. The task force concluded that the conditions which led to his violent and tragically destructive behavior were set in motion early by specific family experiences, which included much physical and emotional abuse, playlessness, paternal over-control, practice with weapons, and other factors, these recurring repeatedly throughout his life. Viewing his last months with the information gleaned from our field interviews and from the autopsy and tissue studies, we felt that the crescendo of drama related to his parents chronically abusive relationship became his perceived responsibility, and was *the* major precipitating stress added to other long term risk factors leading to mass murder. By three months before the Tower tragedy, he was having homicidal fantasies which he had shared with a campus psychiatrist, and though he maintained a Mr. Clean public image, many other evidences of decompensation were occurring. He was receiving daily phone abuse and demeaning diatribes from his father, culminating in a particularly cutting conversation the day before his ascent of the Tower. We interpreted the final violent and suicidal acts as being triggered by his sense of powerlessness, humiliation and entrapment. His inability to find coping techniques through play, humor, safe reciprocal friendships and other distancing and stress-lowering habits were striking findings agreed upon as extremely significant by our team. We had originally expected to discover a brain tumor and drugs as primary causal agents, but our intensive investigation weighted abuse and playlessness as *the* major factors placing him and his future victims at risk.

The Whitman study led to a more organized selected pilot study of 26 young murderers who were interviewed throughout the state of Texas (Brown & Lomax 1969). By design, we chose to examine young males whose only crime had been homicide, and who had been convicted less than two years prior to our study. We matched this group, which was surprisingly culturally heterogeneous, with other young males chosen from similar census tracts. In addition, because a large epidemiological study of medical care in Texas was underway, we were able to cull out a similar group from that activity for comparative study. We interviewed all participants with a structured format, and we found that significant physical abuse had occurred in 90% of these young male murderers. Findings we were *not* expecting were play deprivation and/or major play abnormalities which also occurred at the 90% level. The comparison non-homicidal groups showed abuse and play abnormalities at below the 10% level. Play began to be seen as more important, but exactly how and why was not clear.

Concurrent with this pilot study of young murderers, I was the principle investigator for a multidisciplinary federally funded auto accident research study. Our challenge was to evaluate sequentially, all fatal auto accidents in Harris county, Texas, for a year, looking at as many elements producing the fatality as we could discover, whether induced by the automobile, roadway and weather conditions, and/or behavioral abnormalities of the driver, or drug and alcohol abuse. It was a time of generous funding, so our study was detailed and comprehensive. In addition to behavioral scientists, this investigation involved engineers, a forensic pathologist-toxicologist, an epidemiologist, a radiologist to examine all fractures and their biomechanics and an attorney to help us gain full confidential access to victims and their associates. We were able to involve and had wonderful cooperation from the local police. Our behavioral team used a similar structured interview and questionnaire format to that used in the pilot study of young murderers to obtain our data. As expected, alcohol was involved in 70% of the culpable fatalities. Supportive findings led us to diagnose them as both antisocial personality disorders *and* alcoholics (Bohnert et al. 1968). The 'Drivers who die' group, unlike the murderers and Whitman had not experienced significant childhood physical abuse. Their play histories, however were very similar to the murderers, and very different from the comparison populations.

What *all* of these studies repeatedly revealed and what struck our separate research teams as unexpected, was that (among the other findings) *normal play behavior was virtually absent throughout the lives of highly violent, anti-social men* regardless of demography. Although physical abuse and social (largely paternal) deprivations were significant in predicting chronic risks for violence in the homicide studies, in both the drunken antisocial drivers and the murderers absent or clearly abnormal play was in league with later social and personal tragedy. These were not the findings in the comparison groups. Though they reported many stressful life experiences, their capacities to engage a repertoire of coping capacities appeared related to the richness and variety of play experiences, particularly those early in life. Though these conclusions lacked sufficient numbers and rigorous methodology to become part of established mainstream social-behavioral science, they still remain reasonable, but are unsupported by scientific literature. Women were not studied in sufficient numbers to allow any generalizations about the effects of play deprivation in them, but my remaining impressions are that abuse and play deprivation also take a devastating toll on women. Violence in the small number of homicidal women studied (no comparison groups studied) seemed to be precipitated by abuse and loss, and none we examined were mass-murderers.

Following the human studies on homicide, I wrote to Jane Goodall after reading of the murder–cannibalism by Gombe female chimpanzees Passion and Pom. She responded that they both were ineffectively mothered, and she had observed that their early play and later socialization patterns were constricted. (This exchange of letters led to a friendship and much mutual appreciation of the importance of animal and human play.)

I left clinical research shortly after completing these studies, and became a clinician-administrator in a teaching hospital, curious about and sensitized to the importance of play, but involved more as a traditional clinician with urgent patient and administrative responsibilities. In these capacities, and as a private practitioner over a 20 year duration, I conducted an estimated 8,000 detailed patient interviews. In as many ways as practical, I delved into each patients play history, i.e. their play patterns, imaginative playmates, friendships, pet involvements and toy use, pleasurably repetitive activity of any sort, physical, emotional, musical, fantasy, solo or social. From these anecdotal sources, filtered through the growing information base about how humans develop and socialize, I have gradually come to see play as a separate form of behavior, yet one intricately connected to many other behaviors.

Later in my career, (1986–87) I received a small grant from the Cooper Foundation, and surveyed selected special educational environments and interviewed a few 'genius' scientists, looking for the seeds and harvests of creativity. In such non-pathologic environments, I found playfulness to be the constant companion of kids learning creatively, and it remained active in those scientists whose productivity remained high and varied as they aged.

I also spent much time during 1979–1989 in a variety of film and book productions with the remarkable (and playful) mythologist-scholar Joseph Campbell, and after these were completed, I helped produce a film series for the British Broadcasting Corporation ('Soul of the Universe') which allowed in-depth interviews with some of the world's leading scholars (Brown & Tilby 1992). Thus the earlier years devoted to clinical pathology have been supplemented by these later very different exposures. Play seemed very important as these remarkable personalities achieved mastery, eminence and joy in their adulthood.

I now perceive healthy varied play in childhood as necessary for the development of empathy, social altruism and the possession of a repertoire of social behaviors enabling the player to handle stress, particularly humiliation and powerlessness. I also have found that general well-being and play are partners, and that it accompanies the most gifted in their adult achievements. Perhaps it allows access to the giftedness we all possess.

Over the years, through the medium of clinical evaluation of thousands of patients, helped by the more rigorous methods used studying murderers and clinically antisocial men, I have become convinced that they are not the only ones seriously deleteriously affected by play deficits or abnormalities. A flowing sense of humor and the capacity for engagement in play *is* important for all of us.

Full and free play in childhood also may well serve as an antidote to the development of later poisonous antisocial violent behavior in some, and buffers travail in others. Playfulness is closely allied to other positive factors in the lives of the highly productive and creative. From my clinical practice, I have also seen that a return to play in depressed and stressed adulthood promotes personal healing.

Of course, since play is so integrally linked and seems to 'borrow' from other behaviors, and because rigorous research criteria as to the definition of play are lacking, no specific conclusions about play or its absence in the causality of violence and anti-social behavior was then, or is now, fully warranted. But it appears that we all pay a high price for seriously neglecting it. Playless lives examined in the context of the private practice of clinical psychiatry were often surrounded by states of high accomplishment, but they lacked the exuberance that accompanies a buoyant sense of empowerment or mastery. Depression, over-control, driven ambition, envy and often ecological havoc accompany the play deprived life. In one form or another play starved adults even if economically prosperous were found to live in a narrower, more stereotyped world, and play deprived children are the tragic forerunners of social and personal breakdown.

A period of long reflection following my departure from clinical medicine in 1988 resulted in the decision to explore play broadly as a primary interest. To gain a better grasp on 'the big picture', I spent time as a fellow studying Cosmology under the tutelage of physicist-cosmologist Brian Swimme, and his Center for the Study of the Universe. I wanted to discern how (or if) play could be included in any context within the evolution of complex life. Immersed in cosmological speculations, I began to think of play differently. It could be seen as one of many forms of evolving, emergent, self-organizing, complex dynamic systems. A basic aspect of biologic systems, I learned, is that ever greater information processing has occurred over evolutionary time (Swimme & Berry 1992).

Certain *classes* or *categories* within evolutionary biology take on emergent identities. Creativity emerges in the dynamics of complex systems, and yet, these emergent forms, despite existing on the edge of chaos, are fundamentally *stabilizing* from the lowest to the highest level in the ecological hierarchy (Waldrop 1992; Lewin 1992).

As I have since reflected on the biological implications of current cosmological theory, I see the evolutionary emergence of major new forms of *behavior* such as sleep and play, as fundamentally *stabilizing* complex systems.

Though energized and stimulated by the theoretical linkages between mathematical and computer-generated models of biological systems, I soon realized that the leap from the conceptual to the clinical was too great for direct confirmation. Festering with the need to see authentic play as it manifested itself in the wild, I made the decision to study wild animals at play. I again contacted Jane Goodall. She encouraged me to look up Mary Smith, then a senior editor of the National Geographic

Society. By 1991, I was under contract by them to write an article, consult on and co-produce a TV program on animals at play, adventures which were completed in 1995 (Brown 1994, 1995a).

The experience of worldwide observation of animals in the wild, and time spent with many of the special people who watch them in the field, shaped and enlarged my understanding of play as a broad and remarkable evolutionary phenomenon in all 'smart'* animals. The Geographic sponsorship also allowed me to interview many animal and human play scholars such as Jan Van Hooff in Holland, whose research on the evolution of human smiling and laughing adds wonderful dimensions to an understanding of human playfulness and its ties to the animal world. Another Geographic sponsored interviewee, the well-known ethologist, Iraneus Eibl-Eibesfeldt, had this to say during my 1992 interview with him near Munich.

When I was a young post-war student, disillusioned by the horrors of W.W.II, first beginning my studies in mammalian behavior in 1947 at a small biological field station near Vienna, I raised a badger. It struck me then and I have observed some of the same things time and time again in humans, that during play this badger could become detached from his aggressive and other strong emotions, live in a field free of tension, which allowed him to experiment openly with his own capacities of interaction with the environment. He clearly could set apart his play from his non-play activities. When he was serious, such as when he would get in a real fight, his movement patterns occurred in a very regular order. A set of exclusive non-flexible antagonistic patterns gets activated in a non-play situation. In contrast, when he 'attacked' me in a play-fight, his movement patterns seemed freely independently activated by a different center of control. I think, somehow, this hypothetical center of control allowed this animal to act without anxiety, to freely experiment with new movement patterns. At play, my badger invented new ways to move – he learned to roll down the slope in somersaults and much more. There is no doubt that animals learn many skills relevant to their specific ecology during play. In playing with the badger he often chased me playfully, he often ran away in mock flight. In between, he play-hunted or engaged in playful exploration. Such fast shifts from one behavior to another are never observed in a serious encounter. Once an animal engages in fight or flight, it does not easily change its mood. With the badger, and I have observed similar behaviors in animals and humans, the animals in play act free from the tension which besets them in the non-play situation and this behavior is at the root of what we consider to be the specific freedom of man. We are able to detach ourselves from emotion and obtain freedom to act without anxiety. To consider, to act in fantasy and to perform different strategies in our minds ... a person may be allowed to speak

* 'Smart'. Capable of quick non-stereotypical multiple simultaneous activities: sensory, motor and social.

freely, but his or her utterances may be dogmatic and thus not free at all. For the past twenty-five years I have focused on the cross cultural documentation of human rather than animal everyday behavior, on the rituals and games of children, but my observations about the play of the badger and true freedom remain firmly entrenched in my mind today.

So, as the Geographic projects ended, I had an enlarged view of play which incorporated my early clinical experiences, the cosmology-theoretical possibilities, and the sweeping look at the evolution and variation of play as it occurs throughout nature in 'smart' animals and birds.

I learned that certain evolutionary components have to be in place, for play to occur. First and foremost, a highly complex brain and central nervous system were correlated with the evolution of play. With probably few exceptions, Gordon Burghardt's (1996) playing softshell turtle (see also Burghardt, Chapter 1), for example, play is primarily limited to warm blooded, 'smart' animals.

My opinions about play began to extend beyond clinical guesses to speculations about it's evolutionary roles. I believe its presence can be seen as promoting quickness and flexibility (B. Sutton-Smith pers. commun.). It seems to nullify the rigidity that sets in after successful adaptation. It just may have an engineering role in the evolution of complex behaviors. By now it should be evident that my orientation toward making sense of play has been influenced by many disciplines and personalities. My short immersion in the worlds of cosmology and play ethology had shown me to be an amateur among seasoned professionals. Nonetheless, from my own vantage points, I'd like to try and theoretically generalize from these experiences and in so doing stimulate the reader to ponder play more deeply.

My speculations are that nature (evolution) seems to have provided the brains of playing big brained birds and mammals with an extra 'uncommitted' supply of neurons and connections. This guess is upheld by systematic looks at cortical architecture as its layers exhibit more intricate palisades the more complex the species and by viewing volitional activities. In addition to play behavior, these larger cortices of course reflect the capacity for great cognitive variety. Warm blooded, 'smart' players are not rigidly pre-programmed only for specific consummatory activities, but also respond flexibly to developmental and environmental stimuli. The presence of dynamically interactive circuits in sufficient numbers regardless of their evolutionary trail, seems to be one of the prerequisites for play. Playing creatures have or develop the capacities to receive, integrate, remember and contextualize internal and external signals. They are not always looking for a fight, sex or food, or warily looking over their shoulders for the next higher food-chain representative. Safe and well-fed, they play. How they play, and what constitutes play behavior is becoming less and less controversial as play information accumulates. The external signals that herald play across species lines powerfully affects behavior. Why creatures play, though answered by many plausible speculations, remains as scientifically unanswered as why sleep and dreams exist. Yet play behavior is as fundamental in the lives of players as sleep is in the lives of sleepers. It remains, however, scientifically as enigmatic.

Perhaps play acts like other fundamental super organizers in nature, i.e., it functions as an 'attractor,' a stabilizer, as the matchmaker to newly evolutionarily acquired emergently stabilizing capabilities within complex brains. Perhaps it also offers an alternative to rigidification of patterns thereby enhancing survival. Describing it as such may give play theoretical dignity, but it does so without clinical context, and as I said at the beginning of this primal story of play, 'explaining' clinical phenomena catches my fancy. Eibl-Eibesfeldt describing his badger at play says, '[it] seemed freely [independent], activated by a different center of control.' – such a conclusion is formed from a clinical context.

So, let me offer another example as a means to explore speculatively what may be occurring during an animal play session to amplify my play speculations. Two young lionesses on the Serengeti plain, well fed and safe, approach each other and for the first time begin an exuberant rough and tumble play ballet. They are responding to their intrinsic urges to play. This complex experience seems flexible and free of specific consummatory (more ancient primitive brain stem) demands and thus does not activate the more stereotyped instinctively fixed patterns of predation, dominance or flight-fear which are non-flexible activities. The fun of the experience with its positive freely exchanged affective charge reinforces the internal bodily experiences and the social and sensory outside experiences. This type of play of the young lionesses of this age is repeated again and again in a slightly different choreography, seeming to form one of the basic ways for gaining knowledge of themselves and their environment through adaptive, flexible, and pleasure-laden *action*. Each new ballet seems to add grace and knowledge, awareness, and, may also provide a series of 'value laden scenes' which may form a type of primary animal consciousness (Edelman 1992). During infancy, youth and young adulthood, the brains of all mammals and young birds are particularly responsive to these and many other play activities.

Edelman describes, through elaboration of his *theory of neuronal growth selection,* that perceptual categorization, memory, affect and more, combine to form cortical *maps*, one after another (1992). By becoming sequentially organized into useful adaptive patterns, the lionesses may, in their rough and tumble play example, be establishing new and dynamically interconnected maps each time they play. They are *mutually* developing cooperative non-dominating behavior which seems to have evolved as a unit. The play signals given and received during their mood-altering ballet have meaning to them, are unambiguous, and seem very necessary for their normal development. Yet the interplay between maps, the 'scenes' they contain, are highly individualistic. These maps, the character of which Edelman further elaborates as possessing 're-entrant connectivity,'* remain present and available for modification throughout life, but certainly accumulate most rapidly in infancy and juvenile times. They are, as previously mentioned, also 'value laden' and could certainly form the major cartography of learning. This rough and tumble play by the lioness conveys mood, intent and provides, I am guessing, some of the earliest precursors of social *meaning*. Furthermore, stable socialization patterns are learned during repetition of rough and tumble and other social species-specific acts of playfulness.

It is my surmise from integrating Edelman's thoughts with clinical observation, that play deficient creatures suffer from '*value laden adaptive map deficiency.'* As each map yields new functions, new kinds of memory, and a series of new inner value laden scenes, the player may begin (depending on its evolutionarily derived cartography capabilities) to develop a rudimentary sense of *self-other* (one lioness knows it is not the other lioness; one lioness knows its behavior and play signaling can alter the responses of the other) (Bekoff 1972, 1995; Edelman 1992; Gopnik 1993). Accumulations of life-play experience allow the scene-makers to use the new scene and compare it with previous scenes.

Damasio (1994) believes that the images created (in these scenes) can be volitionally manipulated by higher creatures. Now, leap to humans, add language to this mix and the player has the capacity, in this hypothetical model, to also manipulate language as part and parcel of affect-

* 're-entrant connectivity' which Edelman theorizes is the means by which a dynamically interconnected brain brings percept and concept together in a highly individualistic manner allowing the emergence of primary consciousness in those creatures capable of it and higher order consciousness in human beings.

laden scenes. Edelman's theories, and Damasio's neurologic hunches, when examined through a play lens, allow a plausible ever-more-complex array of *adaptive behaviors* to be conjectured in new contexts.

Jane Goodall, writing in *ReVision* (Goodall 1995) describes chimpanzee symbolic play and joking which seems consistent with a developing rudimentary internal stream-of-consciousness-chimpanzee narrative combined with pretense and symbol manipulation, which, in a game-like situation was shared between the chimpanzee and her caretaker. I can think of similar though less complex episodes while engaged in play and pretense with my labrador retriever, Jason.

Brian Sutton-Smith's (1981) longitudinal cross-cultural observations of the play and games of children and his review of their growing narratives, beginning with fragmented nonsense stories to later privately generated dramas with full structure, gives credence to including narrative as a form of imaginatively created series of interconnected cortical play maps.

Following this trail, I can imagine spontaneous playful linguistic activity as it springs forth in children, and as they socialize, hear their verbal flights begin to take on rudimentary *story* form.

It seems possible then, that narrative mental representations, their 'stories' may comprise a significant part of the child's inner conscious reality. Thus, for me it is not too great a stretch, despite the awareness that there is no single linear path, to follow an evolutionary developmental play trail starting with a 'smart' cortically intricate softshell Nile turtle playfully nosing a ball, to a mountain goat twisting and leaping, to the lionesses rough and tumbling, to chimpanzees playing hide and seek with pretense and deception, to an 8 year old boy making up nonsense stories, to children playing tag, . . ., to a smiling Einstein at his blackboard.

By looking at play as a generator of dynamically integrated affect laden cortical 'maps', of increasing complexity, I believe that play can be considered as a *major* organizer and possible sustainer of our humans dynamic sense of reality.

It seems to me that playless creatures may have an inflexible, narrower, more lizard-like stereotyped sense of 'self' and reality. In a small brained (cortex) cold-blooded reptile, no options for complex cooperative play seem likely. In the murderers previously cited, their inflexibility in the presence of stress, and narrowed repertoire of behavioral responsiveness and enslavement to strong surges of affect could be attributed to 'play map deficiency' from abuse and deprivational circumstances.

Part of the appeal of Edelman's theories and the reason I enjoy incorporating them into play theory is because they seem validated by their illumination of otherwise obscure clinical neurologic syndromes. Oliver Sacks vividly describes some of the strangeness that brain lesions produce, many of which, before Edelman's ideas, lacked a comprehensive theoretical basis (1993). Localized 'map deficits' can explain the findings that occur in brain trauma, stroke, tumor or degenerative diseases, and abuse or deprivationally induced deficits may result in cognitive and socialization dysfunction.

The 'intimate reciprocal relationship in animals as they play', and their capacity to communicate value-laden intentions (what follows is play and will not harm you) so adroitly written of by Marc Bekoff (Bekoff 1972, 1995; Allen & Bekoff 1997), seems in harmony with the theories of the qualia (values) that Edelman writes of in his discussions of animal primary consciousness (Edelman 1992). A blending of field observations and surmises about the cognitive life of, for example Bekoff's coyotes and his dog, Jethro, and the projections from what Edelman postulates is going on inside the brains of animals making 'scenes', I can speculatively unite into an animal symphony with play as the conductor (Bekoff & Allen, Chapter 5).

Sutton-Smith's seminal writings of children involved in games and play and his observations about the efficacy of sequentially more complex group-game play can also be linked to an evolutionary play ladder. He describes children developing the capacity for later *adult-unified behavior* as being dependent upon their successfully mastering such unifying games as ring-around-the rosy (all fall down . . . together) (B. Sutton-Smith pers. commun.). Children's games also teach how to accept authority, learn strategy, teach them how to win, lose, handicap self-or-other in the service of the game, and much more.

Competent workable group adult behavior probably requires competency learned in group play situations in childhood. Thus it seems to me that 'the intimate reciprocal' relationships established as basic and essential in animal play (as in the lioness play example) operate just as importantly in human group play and games and also one on one social play. We too may be the legatees through play of *mutually developing cooperative units,* which aid survival and add joy.

Just as Whitman was not taught the requirements of intimacy and playfulness on an individual basis, the game deprived child may well become the socially dysfunctional adult who cannot handle the complexities inherent in the adult world of corporate and family living.

Despite lacking a solid proven theoretical base, the establishment of playfulness has been effectively used in the business world to enhance well-being, productivity and creativity on the job (Weinstein 1996). Wise parents, gifted teachers and fulfilled adults *live* allowing play a continuous role in their lives.

Conclusion

My personal broad-ranging clinical and speculative review of play over my lifetime may not appeal to the needs of those readers whose satisfactions arise principally through the exercise of academic rigor. It has been my experience, however, that striving to understand the ways in which our own internal narratives and stories have arisen, and how these inner realities intersect with our scholarship and science *is* a means of achieving personal clarity. The stories of play, whether personal or professional, will continue to enchant, beguile and inspire us. Within them lie the energies for discovering more vital meaningful lives.

References

Allen, C. & Bekoff, M. 1997. *Species of Mind: The Philosophy and Biology of Cognitive Ethology*. Cambridge, Massachusetts: MIT Press.

Bekoff, M. 1972. The development of social interaction, play, and metacommunication in mammals: An ethological perspective. *Quarterly Review of Biology*, **47**, 412–34.

Bekoff, M. 1995. Play signals as punctuation, The structure of social play in canids. *Behaviour*, **132**, 419–29.

Bohnert, P. J., Brown, S. L., Smith, J. P. & Pokorny, A. D. 1968a. Drivers Who Die *A Multidisciplinary Report. U.S. Dept. of Transportation,* Grant f-11–6798.

Brown, S. L. 1987, 1988. The Hero's Journey; The World of Joseph Campbell (PBS television program).

Brown, S. L. 1994. Animals at Play. *National Geographic Magazine*. **186**, 2–35.

Brown, S. L. 1995. Through the Lens of Play. *ReVision*. Spring. **17**, 2–12, 35–42.

Brown, S. L. & Cousineau, P. 1990. *The Hero's Journey*. San Francisco: Harper.

Brown, S. L. & Freedman, D. A. 1968b. On the Role of Coensthetic Stimulation in the Evolution of Psychic Structure. *Psychoanalytic Quarterly*, **37**, 418–38.

Brown, S. L. & Freedman, D. A. 1970. A multi-handicapped baby: the first eighteen months. *Journal of the American Academy of child Psychiatry*, **9**, 19–30.

Brown, S. L. & Lomax, J. 1969. A pilot study of young murderers. *Hogg Foundation, annual report*. Austin, Texas.

Brown, S. L. & Moses, C. 1995. Play, the Nature of the Game. *National Geographic Explorer* (TBS television program).

Brown, S. L. & Tilby, A. 1992. The Soul of the Universe (BBC television program).
Brownlee, S. 1997. Play: It's not just fooling around. *U.S. News and World Report*. 3 February, pp. 45–8.
Burghardt, G. 1996. Problem of reptile play; environmental enrichment and play behavior in a captive Nile softshell turtle, *Troinyx triunguis*. *Zoo Biology* **15**, 223–38.
California Academy of Sciences. 1992. Symposium: Neoteny and the Evolution of the Human. San Francisco.
Damasio, A. 1994. *Des Cartes Error* . Avon Books, New York. p. 18–113.
Edelman, G. 1992. *Bright Air, Bright Fire*. New York: Basic Books, Ch. 10,11, pp. 118–21.
Goodall, J. 1995. Chimpanzees and others at play. *ReVision*. Spring Edition **12**, 14–20.
Gopnik, A. 1993. Psychopsychology. *Consciousness and Cognition*, **2**, p. 264– 280.
Lewin, R. 1992. *Life at the Edge of Chaos Complexity*. New York: Macmillan.
McMahon, F. & Sutton-Smith, B. 1995. The past in the present: Theoretical directions for children's folklore. *Children's Folklore: A Sourcebook*. (eds. B. Sutton-Smith, J. Mechling, T. Johnson, and F. McMahon). New York: Garland.
Meeker, J. 1997. *The Comedy of Survival*. University of Arizona Press.
Pellegrini, A. D. 1989. Elementary school children's rough-and-tumble play. *Early Childhood Research Quarterly*, **4**, 245–60.
Sacks, O. 1993. Making Up the Mind. *New York Review of Books*. pp. 42–49.
Sutton-Smith, B. 1972. *The Folkgames of Children*. Austin: University of Texas Press.
Sutton-Smith, B. 1974. *The Anthropology of Play*. Association for the Anthropological Study of Play, **2**, 8–12.
Sutton-Smith, B. 1981. *The Folkstories of Children*. Philadelphia: University of Pennsylvania Press.
Sutton-Smith, B. 1993. Dilemmas in adult play with children. In K. McDonald (ed.), *Parent-Child Play Descriptions and Implications*. Albany, New York: State University of New York Press.
Swimme, B. & Berry, T. 1992. *The Universe Story*. New York: Harper Collins.
Waldrop, M. M. 1992. *Complexity*. Touchstone Books. pp. 9–13
Weinstein, M. 1996. *Managing to Have Fun* Fireside Books: New York: Simon and Schuster.

Index

Note: page numbers in *italics* refer to figures and tables